ALIEN
INVASION

ALIEN
INVASION

AMERICA'S BATTLE WITH NON-NATIVE ANIMALS AND PLANTS

ROBERT DEVINE

NATIONAL
GEOGRAPHIC
SOCIETY

To Mary and Sarah
For support beyond measure

Published by the National Geographic Society
1145 17th Street, N.W.
Washington, D.C. 20036

First printing, June 1998
Copyright ©1998 Robert Devine. All Rights Reserved.

Library of Congress Cataloging-in-Publication Data:

Devine, Robert, 1951-
     Alien invasion : America's battle with non-native animals and
plants / by Robert Devine.
         p.   cm.
     Includes index.
     ISBN 0-7922-7372-9
     1. Biological invasions—United States. I. National Geographic
Society (U.S.) II. Title.
     QH353.D48 1998
     577'.18—dc21   98-9730
                                                              CIP

Printed in the United States of America

CONTENTS

FOREWORD
BY BRUCE BABBITT

America is being invaded. Alien species—plants and animals outside their natural ranges—are spreading throughout our lands and waters. Some scientists call this invasion a biological wildfire, others label it biological pollution. Whatever the term, it's a growing problem we need to address.

Those of us concerned about the health of America's public lands have a particular interest in the alien invasion. Many of our finest natural areas have suffered terribly at the hands of these invaders. In Everglades National Park a non-native shrub called Brazilian pepper has infested tens of thousands of acres; in places it has formed dense tangles in which few native plants and animals can survive. In Idaho's Snake River Birds of Prey National Conservation Area, an ecological chain reaction set off by an alien grass threatens populations of golden eagles and prairie falcons. A lovely exotic plant called purple loosestrife belies its beauty by degrading wetlands in many of our national wildlife refuges, to the detriment of native waterfowl and many other species. A whole complex of aliens—including pigs, rats, and mosquitoes—have contributed heavily to the extinction of several native birds that once graced Hawaii Volcanoes National Park.

But this is not just a problem for agencies that manage public lands. The alien invasion is transforming the entire American landscape. It affects us all.

When populations of native animals and plants decline, even become extinct, we lose some of our biological diversity, that wondrous variety of species forming our most valuable natural heritage. Each species is precious in its own right, but together they create the basis for all life, including human life. These millions of species, from grizzlies to grass, provide vital services. Various plant, animal, and microbe species purify our water, produce the fish we catch, turn dead leaves into nutrients, create medicines that cure our illnesses, control floods, and help regulate the very atmosphere that makes life as we know it possible. Aliens deplete our native biodiversity, creating diminished habitats that harbor only a handful of plant and animal species where there used to be dozens or hundreds. Exotic invaders have played a significant role in the listing of nearly half of all federal threatened and endangered species.

Non-native plants and animals also cause substantial economic harm. Aquatic weeds choke irrigation canals and flood-control channels. Leafy spurge ruins ranchlands. Zebra mussels clog utility pipes. Hordes of exotic insects attack crops. Predatory alien fish decimate native fisheries. In terms of expenses for control and lost productivity, the alien invasion costs businesses, consumers, and taxpayers billions of dollars every year.

Because the invasion diminishes the life of every American, it should concern every American. We need to work together to deal with this issue. If one county works hard to control an invading weed but neighboring counties do little, the weed will spread and remain a constant problem. On the other hand, if counties coordinate their efforts with one another, and with landowners, citizen groups, businesses, and other levels of government, they can contain the weed, perhaps even eradicate it. Creating robust partnerships is one of the goals of an initiative on invasive exotic species that the federal government has recently launched.

The good news is that we have the means to cope with the alien species. The invasion is still at a fairly early stage. If we act quickly, decisively, and in concert with others, we can keep our biological systems healthy and our quality of life intact. In *Alien Invasion,* you'll read about the successful efforts of scientists, business owners, farmers, ranchers, citizens, government agents, and industry executives to halt this expansion. But such efforts are still not common enough, largely because most Americans are unaware of the invasion and its potential consequences. Raising the level of awareness about this issue is perhaps the greatest service *Alien Invasion* can render. As the first nontechnical book about the invasion, it introduces the general public and policy makers alike to an issue that none of us can afford to overlook.

PART I: INVASION

CHAPTER 1

INVADERS FROM EARTH:

AN INTRODUCTION

About 15 years ago I took a weekend birdwatching class in Colorado. The teacher was a 30-something woman who brimmed with a gentle passion for birds and for all the natural world. As we roamed the foothills near Denver one fresh spring morning, she gushed over sparrows and finches and told us in loving detail about the quirks of hummingbirds and hawks. At one point, gesturing toward a flock of chattering birds that flashed iridescent greens and purples, I asked her what they were. European starlings, she said. Eyeing them coldly, she said that she keeps a BB gun at home to shoot any starling that happens into her yard.

Taken aback, I asked her why she of all people would shoot birds. Not birds in general, she answered, horrified at the thought. Only starlings, an invasive alien species. She said she hated to kill anything, but she felt compelled to act because starlings usurp habitat and resources from native birds, including some declining species. She could have added that starlings are a costly public nuisance, but it was concern for her beloved native birds that motivated her to pull the trigger of her old Daisy air rifle.

This conversation served as my introduction to the alien invasion.

Starlings are a prime example of an invader. Unfortunately many such examples exist in the United States; fortunately there are better ways to combat alien species than to plink at them with BB guns. But before we can subdue these invaders, we must first understand the problem.

The alien invasion is the least known of the world's major environmental issues. Some of our ignorance regarding these intruders stems from the fact that often they come wrapped in pretty packages. Our hearts quicken when we see a band of wild horses galloping along a distant ridge, and we savor the showy blooms of purple loosestrife waltzing with the breeze. Most of us don't realize that wild horses degrade native grasslands and purple loosestrife destroys wetlands. Even despised alien species seldom alert us to the alien invasion because we don't realize that they're aliens. Crabgrass, most urban rats, carp, and dandelions are among the many foreign pests that have been around so long that we think of them as a natural part of the landscape. In many cases, we overlook the harm caused by alien organisms because it is subtle, long-term, or occurs out of sight. We don't see salt cedars monopolizing scarce desert water or the mites that have killed nearly all of our nation's wild honey bees.

People who know the most about alien invaders are the most worried. A rapidly growing number of scientists and land managers rank the invasion as one of the world's and the nation's most serious environmental problems. Edward O. Wilson, the Pulitzer Prize-winning Harvard entomologist, believes that aliens are second only to habitat loss in depleting biodiversity, and his ranking blurs because aliens are one of the main causes of habitat loss. When the National Park Service examined 101 conservation issues that afflict our park lands, they found that exotics were the most widespread problem.

In 1997, more than 500 scientists, agricultural officials, conservationists, and land managers sent a letter to Vice President Al Gore asking that the federal government take concerted action against alien species. "Biological invasions produce severe, often irreversible impacts on agriculture, recreation, and our natural resources," the letter states. "In some instances, they even have major human-health consequences. The 21st century holds the clear threat of further devastating invasions unless a coordinated national effort is established."

As the letter notes, alien species hurt the economy as well as the environment. Ranchers, farmers, the tourism industry, commercial fishers, utility companies, the timber industry—many businesses lose a great deal of money to aliens. The Weed Science Society of America, for example, estimates that invasive weeds cost American farmers about $4 billion a year. An alien insect called the cotton boll weevil has cost the cotton industry an estimated $13 billion. Such losses in turn cost American consumers and taxpayers a great deal.

In addition to harming the environment and the economy on a grand scale, alien species diminish our quality of life in a host of little ways. Yellow starthistle, a spiky alien weed, blocks off hiking trails. Some of the cold viruses that lay you low in the winter originated on another continent. Mute swans, an aggressive European species, attack any angler who strays into their territory. And let us not forget dandelions, scourge of lawns across the nation. A homeowner whose back is stiff from digging out dandelions has experienced in a small way the harm inflicted by invaders from Earth.

Aliens: a.k.a. non-native, exotic, introduced, or non-indigenous species. These labels apply to any animal, plant, or microbe found outside its natural range. Political boundaries mean nothing. A species found in the northern part of a county in North Carolina is considered exotic if it doesn't naturally occur there, whether it came from South Korea, South Carolina, or from the south end of the county.

It is believed that hunting enthusiasts from the state of Washington captured mountain goats in British Columbia and Alaska, where this species is native, and took them to the Olympic Mountains in northwest Washington, which mountain goats had never reached due to the range's geological isolation. Visitors admiring mountain goats in Olympic National Park are looking at exotics, and visitors admiring mountain goats a hundred miles away in British Columbia are looking at natives. Scientific hair-splitting? Not at all. The mountain goats in British Columbia fit right into their ecosystem, but their brethren in the Olympic Mountains, animals that don't belong there, damage the habitat: Native plants in Olympic didn't evolve under pressure from mountain goats and therefore don't stand up well to their grazing and trampling. Though non-native species may come from the other side of the mountain, most come from farther away, usually from another continent.

Virtually all invasive species are non-native, but not all non-natives are invasive. In fact, many exotic species don't cause trouble in their new environments. The presence in the U.S. of irises, though they're of foreign origin, shouldn't keep us awake at night. They behave demurely in their adopted homeland and lend their beauty to our gardens. Many other exotic species also provide benefits to our society. Wheat, soybeans, oranges, tomatoes, rice, apples, and almost all the rest of our major grains, fruits, and vegetables were derived from plants native to other continents. Many other aliens are neutral, doing neither good nor harm. It's the invasive ones that we have to watch out for, the ones that proliferate out of control, degrade our ecosystems, make us ill, and devour our crops.

It's difficult to predict which non-indigenous species will be invasive. Some of the aliens currently deemed innocuous or desirable will turn out to be trouble. For nearly 50 uneventful years, Florida gardeners adorned their yards with Asian fig plants. A few years ago the figs suddenly became invasive, spreading through gardens and into natural areas. What turned Dr. Jekyll into Mr. Hyde? The fig's natural pollinator, an Asian wasp, finally had followed its host plant to the United States.

Numerous introduced species follow the Asian fig's pattern, but usually the reason for the change from shrinking violet to aggressive invader isn't as clear. In fact, a critical question in the study of exotics is why some organisms run amok when moved to new lands, whether right away or decades later. Melaleuca trees don't rock the boat in their native Australia, yet they are ravaging South Florida.

To some degree, organisms are held in check in their native habitats by their native predators and competitors, which have evolved ways to prey on and compete with those organisms. Beyond their natural ranges, free from familiar constraints, exotic organisms are primed to run wild. Invasion is more likely, too, if an alien settles in a place that features a climate similar to that of its homeland.

In many cases invasiveness also depends to a degree on the health of the exotic's new habitat. A disturbed area, such as a clearcut forest or a dammed river, generally is more vulnerable to invasion than is an intact ecosystem. Certain traits also increase the likelihood that a species will be a successful invader: for example, the ability to reproduce prolifically, an eclectic diet, the ability to spread rapidly, and hardiness in the face of harsh weather. Watching for these and other characteristics helps people predict invasiveness, but it is not an exact science.

The number of non-native species that do cause harm remains hazy; until the last decade or so, the scientific community largely neglected exotics. Little had been done to pull together information about introduced species in the United States until 1993, when the Congressional Office of Technology Assessment (OTA) published the landmark study *Harmful Non-indigenous Species in the United States*. The 391-page report includes many provocative numbers: for example, the OTA counted 4,500 non-native species that have become established in the U.S. as "free-living populations beyond human cultivation or control." The OTA also tried to determine how many non-indigenous species make trouble. For example, the report figured that 30 percent of all exotic fish species create problems, 28 percent provide benefits, 17 percent do both, and the effects of 25 percent are neutral or unknown. Overall, about 15 percent of all exotics in the nation were found to "cause severe harm."

Harm comes in many shapes. Sometimes it's a simple matter of an exotic eating natives, such as the cats that pounce on millions of native birds every year. Most of these birds aren't candidates for extinction—at least not yet—but predation by the wild cats diminishes their numbers and alters the species mix within certain natural communities.

Other times the aliens displace natives by outcompeting them. Cheatgrass takes water from the soil before the roots of native bunchgrasses can reach the moisture, resulting in vast expanses of cheatgrass in the Great Basin, where bunchgrasses and dozens of associated native plant species once grew. Some exotics disrupt the food web, producing a ripple effect that changes the whole community. In some lakes and rivers in Glacier National Park, an introduced shrimp species now uses most of the resources that once sustained native zooplankton. As zooplankton populations have declined, so have the populations of forage fish. That in turn has starved or driven away eagles, otters, bears, and other predators that relied on the fish.

Researchers and land managers also worry about the changes in ecosystem functions caused by some exotics. Salt cedar for one. This alien tree overruns the banks of rivers, creeks, and wetlands in the Southwest. Salt cedar sucks up and transpires water at a rate far exceeding that of the native vegetation it displaces, drying out critical desert water sources. Stanford biologist Peter Vitousek explains that an alien in this category "would not merely compete with or consume native species—it would alter the fundamental rules of existence for all organisms in the area."

By harming native plants and animals, alien invaders diminish biodiversity—that wondrous variety of living organisms that is crucial to the health of any ecosystem. The loss of biodiversity is especially vivid when exotics drive a native species to extinction, but biodiversity also suffers when there's too big a drop in a species' population, even if the species doesn't go extinct. When the exotic fungus known as chestnut blight killed a billion American chestnut trees in the early 1900s, it depleted North America's biodiversity, even though a handful of trees survived.

The demise of the American chestnut provides a simple lesson in the practical value of biodiversity—valuable wood that had special qualities prized by builders was destroyed. The rosy periwinkle provides another example. This plant produces alkaloids that cure most people who suffer from Hodgkin's disease or from a lethal form of leukemia. And in the 1970s, a plant disease called coffee rust roared into Central and South America, threatening the coffee plantations. Searchers tracked down wild coffee trees in Ethiopia that harbored a resistant gene, which scientists bred into the cultivated coffee trees, saving the plantations.

The practical value of biodiversity to humankind runs deeper than coffee; the primary function of a plant or animal is not to supply us with commodities but to play its role in a dynamic and complex ecosystem. All those intricately functioning organisms produce what scientists refer to as ecosystem services. These include pollination, water purification, pest control, the creation of soil, and other fundamental benefits. The diverse species that together carry out these functions exist everywhere, in managed landscapes as well as in the wilds. Nitrogen-fixing microbes in the soil of our nation's farmlands, for example, supply some $7 billion worth of this essential element every year. Simply put, if biodiversity suffers too much, if we lose too many species in too many places, then the ecosystems that sustain us will unravel.

Imagine a seed. Let's say a seed from a Chinese tallow tree. Though Chinese tallows are native to several Asian countries, they now infest nine southern states. How might this invasive pest have travelled to the United States?

A typhoon might have whisked the seed into the South China Sea, where it may have landed on a floating log and drifted 8,000 miles to the California coast, where a beachcombing raccoon might have eaten it and deposited it in a suitable—and well fertilized—site. Over the course of millions of years this could have happen, but it's hardly a likely event.

Perhaps the seed became lodged in the feathers of a bird, some far-ranging migrant that flew over the sea to the U.S. But extremely few birds migrate that far.

Or it may have travelled the overland route, edging a little farther west with each generation. Again, unlikely. There would have been serious barriers to cross—the kind that define the boundaries of a species' native range, such as the Himalaya Mountains, the Gobi Desert, and the Atlantic Ocean. . . .

In the natural order of things, alien species can and do get past oceans, mountains, deserts, and other such barriers—but rarely. Very rarely. And when an exotic did somehow surmount these obstacles, it usually failed to establish its species, shut out by a robust ecosystem in which physical space and ecological niches were filled. And when it did become established, that pristine system likely kept the intruder in check. Exotics always have been a part of nature's dynamics, but an extremely minor part.

Humans changed those dynamics, particularly with the development of trade and agriculture. Trade hastened the exchange of animals, plants, and microbes among regions. Farming and livestock grazing disturbed natural ecosystems, creating the openings for non-indigenous species. From about 10,000 years ago to about A.D. 1500, the natural trickle of aliens grew into a stream.

Then came the torrent. Explorers, traders, and settlers used improved means of transportation to rapidly shrink the globe, greatly increasing both the intentional and accidental exchange of organisms. The Industrial Revolution and the accompanying explosion of human population and industrial-strength agriculture disrupted the majority of the world's natural places, leaving them vulnerable to exotics. Humankind's acceleration of the movement of species around the planet prompted the eminent British ecologist Charles Elton to write: "We must make no mistake: we are seeing one of the great historical convulsions in the world's fauna and flora."

But most people do not see this acceleration. Invasion usually is a slow process compared with the drama of oil spills and clearcutting. Yet by nature's standards, the pace of introduction has become extremely rapid: hundreds and thousands of times faster than is natural. Consider Hawaii, whose isolation made it exceptionally difficult to reach by natural means. Researchers calculate that prior to the arrival of humans, a new species arrived on this chain of islands about once every 70,000 years. Today, despite efforts to keep non-natives out, a new exotic becomes established every 18 days. That's about a million and a half times faster than the natural rate.

Non-native species come to the United States in a variety of ways. Many arrive by accident. The OTA report lists some classic examples. The notorious zebra mussel stowed away in the ballast water of a ship and was pumped out into the Great Lakes. Serrated tussock, a noxious range weed, slipped into the U.S. from Argentina via shipments of contaminated pasture and lawn-grass seeds, sold through WalMart, K-mart, and Ace Hardware. The Asian tiger mosquito, a potential carrier of several deadly diseases, entered the country in a containerized shipment of used tires. Hiking boots, military planes, car tires, ships' hulls, packing materials, cut flowers: the paths that accidental invaders take to the United States are virtually infinite.

Many species arrive in ways that might be described as semi-accidental. Typically someone imports a species with the intention of keeping it confined forever, but somehow the species escapes. Africanized honey bees (a.k.a. killer bees) escaped from a research facility; Andean pampas grass, a weed that invades natural areas, sneaked out of someone's garden. Exotic pets regularly escape. Some, such as African giant snails (the size of a fist), monk parakeets, and at least 27 species of non-native aquarium fish have established free-living populations in the U.S. From 1988 to 1992, the Massachusetts Division of Fisheries and Wildlife reported the following pet escapes: a Nile monitor lizard, a bobcat, three boa constrictors, hundreds of tropical birds, three wallabies, nine European fallow deer, and a crocodile.

Then there are the non-natives that are brought into the U.S. intentionally. Though this isn't quite the free-for-all it used to be, individuals, businesses, and the government still import plenty of non-native species that they think might be useful. Some procurers of exotics are utterly unaware of the potential for harm, and some are willfully ignorant.

Take Chinese tallow. Benjamin Franklin brought us this pernicious pest. In the late 1700s, he shipped some tallows to a friend in South Carolina to cultivate for candle wax; for more than a thousand years they'd been prized in their native range in eastern Asia for their oily seeds. Franklin's actions made sense in the 18th century; the word "ecology" hadn't even been invented yet. He learned of something that might be useful, so he acquired it for his friends and country. Unfortunately Chinese tallow went on to invade the South where it overruns bottomland forests and wet prairies.

The award for sheer productivity goes to the U.S. Department of Agriculture. In its early years, its scouts zealously combed foreign lands; by 1923 the department had introduced more than 50,000 exotic plants into the United States. Unfortunately some of these species became major invaders, such as crabgrass (introduced as a grain), water hyacinth, and Johnson grass. In addition, hiding on those thousands of plants were many of the insect pests and plant diseases that give today's farmers headaches.

Other government agencies followed the Agriculture Department's lead. In an effort to curtail erosion, the Soil Conservation Service, the Army Corps of Engineers, and state highway departments brought in some plants that have turned out to be among our nation's worst invasives, including kudzu (dubbed "the vine that ate the South"), European beach grass, and multiflora rose. State fish and game departments and their constituents leaped at the chance to add to what nature had granted North America in the way of quarry for sport hunting and fishing. They introduced axis deer, brown trout, European boar, peacock cichlid, Barbary sheep, Hungarian partridge, and dozens of other species. Fisheries managers also enthusiastically transplanted native North American fish to regions in which they were foreign—striped bass and rainbow trout weren't always found everywhere. In the late 1800s, specially designed rail cars hauled eastern fish west and western fish east. Contemporary accounts tell of workers randomly dumping fish from bridges into the rivers and creeks below while the steam engines paused to take on water. In recent years, government agencies have tempered their passion for tampering with nature, but they still tinker more than a little.

That old nemesis, the European starling, illustrates the truly daunting diversity of rationales that people used to intentionally introduce exotics to the U.S. The seeds of the starling invasion were sown, forsooth, by William Shakespeare, though responsibility actually rests with the American Acclimatization Society. Acclimatization societies were all the rage in the 19th century, when the United States teemed with recent immigrants who pined for the familiar plants and animals they'd left in the old country. Infused with the can-do spirit of their new country, these societies recruited species from their members' homelands. The American Acclimatization Society's peculiar ambition stemmed from the dual passions of its leader, Eugene Scheifflin, a drug manufacturer and resident of New York City. Scheifflin loved birds and Shakespeare, so he set his society to the task of bringing to America every bird species mentioned in the works of the Bard. Unfortunately Hotspur says in *Henry IV*, "Nay, I'll have a starling shall be taught to speak nothing but 'Mortimer'. . . ." So in 1890, the society released a few starlings in Central Park. Now hundreds of millions live throughout the country, making starlings the most numerous species of bird in the nation.

How did starlings expand from a few to hundreds of millions? That's the other half of the invasion equation—the spread within the United States. If a few starlings had taken up residence in Central Park, their presence a source of pleasure to Scheifflin and his fellows, no harm would have been done. An introduction alone is not a problem. It's just the beachhead. It's the ensuing invasion that wreaks havoc.

In August 1997, I received the monthly newsletter published by the local government of the small Oregon city where I live. A one-page article informed residents that earlier that summer a lethal fungus, Dutch elm disease, had infected five trees within city limits: the first appearance of Dutch elm disease (DED) in our town. The story went on to describe the policy for dealing with DED. Our City Council adopted this policy in 1987, after the disease had shown up in a city

80 miles away; the council knew it was just a matter of time before DED reached us. Dutch elm disease has been rolling across the nation since 1931, when it entered the United States at an East Coast port in a shipment of elm wood from France. The body count is some 77 million elms and climbing.

The ability of invasive species to spread and persist sets them apart from most other environmental problems. If a developer drains a wetland to build a shopping mall, it's a significant loss. But that shopping mall won't of its own accord expand and cover other wetlands. Nor will birds carry mall seeds 100 miles and sprout new malls. Similarly, when a supertanker spills oil into the ocean, that oil degrades the habitat for years to come, but it can't produce more oil. Invasive species, on the other hand, are alive. They reproduce, so they'll be around indefinitely.

Fortunately we can do much to stifle the alien invasion. Some solutions are easy, such as not releasing our unwanted aquarium fish and plants into the environment. Some require hard work, such as trapping wild pigs, pulling Scotch broom, and bulldozing Brazilian pepper. Some solutions make use of advanced science, such as developing American elms that rebuff Dutch elm disease and boosting populations of native insect predators that prey on crop pests. Some require changes in government policy, such as energizing the feeble laws regulating noxious weeds and putting a few dollars into control programs. Some depend on business; for example, the nursery industry needs to quit importing and selling invasive plants. Perhaps most important, and most challenging, we need to change some of our habits. Those of us who garden will have to think twice about what we plant in our yards, and those of us who fish will have to stop sowing our favorite game fish in every body of water in the country.

If we vigorously pursue the full range of solutions, what results can we expect? Certainly not the eradication of all the invasive exotics in the U.S. A range manager once told me that "there's not enough money in the U.S. Treasury to get cheatgrass out of everywhere," and that's just one species. But that doesn't mean the only option is to wring our hands in despair. We can eliminate some invasives. We can control those we can't eradicate, minimizing the damage they cause. We can keep many potential invaders out of the country so that we don't have to look over our shoulders the whole time we're working on the problems at hand. And we can care for the land in ways that make it more resistant to invasion.

Whatever we do, we can't afford to keep ignoring or underestimating the invasion. The environmental and economic harm already is serious, and it will worsen rapidly unless we mount a substantial response. We also need to think about less obvious but no less important consequences. Left unchecked, non-indigenous species eventually could transform much of the American landscape, which in turn would transform the way many of us live. Already, invasive weeds have forced people off their ranches, fire ants have made parks and back-yards unsafe for children, water hyacinth has covered favorite fishing lakes with impenetrable tangles of vegetation, and avian malaria has joined other exotics in driving several species of Hawaiian birds to extinction. Take these examples, add all the other such problems caused by invasive species in the U.S., and you get scattered but significant changes in our way of life. Now think about multiplying those problems by 10, by 50, by 100 as the species already here expand and new species arrive. Though the alien invasion already is well underway, it has reached only a fraction of its full potential.

Consider a disaster that almost happened and surely will happen if we're not vigilant. In 1991, a task force of scientists, economists, and government officials calculated the impact of the invasive species that likely would enter the United States if some Northwest lumber mills imported logs from Siberia. They looked especially hard at the Asian gypsy moth and the nun moth, two Siberian species that can devastate forests if moved beyond the controls of their native habitat. The task force estimated that between 1990 and 2040 these two species would inflict losses of $35 to $58 billion on the Northwest. As it turned out, their calculations were almost tested before the ink was dry. While the task force was pondering this problem, Asian gypsy moths entered the Northwest in grain ships. An emergency control program apparently killed them all, though it's possible that some moths escaped and are out there in the forests somewhere breeding.

Given the enormous dollar cost of the two Siberian moths, imagine the extent of the less tangible costs. The community fabric of some timber-dependent towns would be ripped apart: friends and family scattered, stores vacated, playgrounds quiet. Think of Mt. Rainier National Park, Olympic National Park, and the Northwest's many other national parks, state parks, national forests, and wilderness areas, and picture them riddled with defoliated wastelands. Camping, hiking, and Sunday drives in the Northwest would have lost much of their appeal. Consider the harm that would have been done to the flora and fauna of the region. Widespread damage to the conifers, which anchor many Northwest ecosystems, would have led to the decline of elk, mushrooms, salmon, lichens, spotted owls, and hundreds of other forest species. We in the Northwest would even have lost some of our identity, our sense of who we are, because our regional culture is intimately tied to our grand conifer forests.

The alien invasion sets up a classic penny-wise and pound-foolish situation. Heed the sorry tale of the Brazilian pepper tree. In 1957, an Everglades National Park biologist noticed a Brazilian pepper growing alongside a park road. One tree, no problem, he noted in a letter to the park superintendent. Today Brazilian pepper has spread to 100,000 acres of the park, crowding out many native plants and the wildlife that rely on them. Park managers are spending tens of millions of dollars to remove this invader from just one key 4,000-acre tract. Nobody even knows how to control pepper out in the mangroves. If only the park staff had known to kill that first plant and to keep killing any Brazilian pepper that cropped up.

Today we know. We understand that for a penny—well, maybe a few pennies—we can prevent and cut short invasions that will cost a pound if we procrastinate. How many times have we contemplated overwhelming environmental problems and wished that we'd recognized and dealt with them in an earlier stage? What if we had started protecting old-growth forest when 80 percent of it remained instead of when 80 percent of it was gone?

The alien invasion is in that early stage. Certainly, it's already a big problem, but not nearly as big as it will be if we don't contain it. And we can contain it. We have the means, both in terms of knowledge and resources. That gives us the rare opportunity to handle a major environmental problem well before it becomes overwhelming.

We can do a lot without even lifting a finger. Gardeners can learn which plants are invasive in their areas and simply not plant them. Airline passengers returning to the U.S. from overseas can fill out their declaration forms honestly and completely, so they don't inadvertently bring in some invasive pest. Pet owners can refrain from releasing unwanted tropical fish or exotic birds into the wild. We can accomplish much by simply becoming more aware.

Achieving that awareness presents a challenge; the reality of the invasion often runs counter to our perceptions. How can a beautiful plant or a sweet-singing bird do more damage than a toxic waste dump or a developer's army of bulldozers? It wasn't that long ago that I didn't understand the full import of the invasion myself, even though I've been writing about environmental issues for many years. I encountered that starling-slayer in Colorado 15 years ago, but I didn't appreciate the magnitude and urgency of the situation until nearly ten years later, when I took an eye-opening journey through a part of the West that has been devastated by exotic weeds. That's when I began to realize that the alien invasion has the potential to sicken our land, to harm the marvelously diverse plants and animals whose fates we hold in our hands, and to leave for our children and their children and all succeeding generations a greatly diminished world.

CHAPTER 2

PIG TROUBLE, U.S.A.

The person wearing the pink pig costume stood out among the three-piece suits that plied the block of high-rise office buildings in Arlington, Virginia. It was lunchtime on a weekday in 1994, and hungry workers streamed past the pig on the sidewalk. As the pig and a cohort of about a hundred uncostumed companions raised banners and posters in front of national headquarters for the Nature Conservancy, their mission became evident: THE NATURE CONSERVANCY—TORTURING ANIMALS WITH YOUR MONEY and THE NATURE CONSERVANCY KILLS KIDS. The group was staging a protest against the Conservancy's snaring of non-native pigs and goats on its preserves in Hawaii.

While waving banners and posters and brandishing pig skulls, the protesters chanted, "TNC, TNC, shame on you, shame on you," and the like. Trying to sow outrage and garner support, they buttonholed passersby and told them of the suffering of snared animals. Having been tipped off, the police, some reporters, and a few nervous Conservancy staffers looked on. The presence of the police was crucial to the protestors because public arrests were part of the game plan.

After shouting awhile, the costumed pig and several other protestors charged Conservancy headquarters. Perhaps they intended to chain themselves to the doors, as protestors had done at other Conservancy offices, but the forewarned TNC staff had removed the door handles and locked the front doors and every other entrance to their headquarters. The frustrated protestors began banging on the glass doors. The police officers moved in to pull them away. In the scuffle a Conservancy staffer was knocked to the ground. Then someone swung a bucket. A wave of blood-red liquid splashed against the front doors, splattering Conservancy employees and others in the vicinity. Still unarrested, nine of the determined protestors filed out to the busy street and lay down, stopping all traffic. The police played their role in the civil-disobedience game plan and arrested all nine of them. The blood-red liquid turned out to be paint.

Other costumed pigs appeared around the country in 1993 and 1994. The confrontation at TNC headquarters in Virginia was only one of several protests in an intensive campaign against the Conservancy waged by People for the Ethical Treatment of Animals (PETA), an animal-rights organization claiming 600,000 members. Joined by like-minded animal activists, PETA showed up at other Conservancy offices and TNC annual meetings. In one case, Alex Pacheco, cofounder and current PETA chairperson, hid backstage with another PETA member while the chairman of TNC's Hawaii chapter gave a speech. Suddenly Pacheco and associate burst onto the stage, interrupting the speech with a blast from an air horn taped to Pacheco's wrist. Seizing a microphone, Pacheco gave a ranting speech, addressing the audience about the suffering of pigs and goats and TNC's role—until security hauled him and his companion away. Sometimes the protests got even more personal. An Easter basket addressed and delivered to John Sawhill, president and CEO of the Conservancy, contained a stuffed toy rabbit covered with ersatz blood, probably paint. Occasionally TNC staff members received threatens over the phone. PETA disavows such personal tactics.

PETA didn't limit its campaign to Conservancy staff and officers. Heeding famous advice given by the Watergate informant, "Deep Throat," to follow the money, PETA went after TNC members, donors, and corporate supporters. Among corporate sponsors, the Nature Company took the most heat. Actions included a national boycott, protestors chaining themselves to the doors of Nature

Company stores and surreptitiously slipping scathing literature into books for sale at those stores. As in the case of the Conservancy, Nature Company staff and officers also received phone threats.

Conflict between a conservation group and an animal-welfare group at first seems strange; animal-rights advocates and conservationists share many of the same goals regarding wildlife. Before PETA's pink-pig and paint campaign, some PETA staffers belonged to the Nature Conservancy and vice versa. PETA and TNC had even exchanged mailing lists, assuming the members of each organization to be of like mind on many issues.

So why didn't the Nature Conservancy give in and stop snaring pigs? Conservancy staffers and members shared PETA's revulsion for the painful deaths that snaring sometimes inflicts on wild pigs. PETA's campaign was costing the Conservancy a great deal of time and money. Conservancy staffers were being harassed by animal activists and were grilled by upset Conservancy members. Why endure all that misery?

Because the Nature Conservancy is just as passionate about protecting the planet's ecological well-being as PETA is about protecting animals from pain and suffering. And to save the embattled ecosystem, Conservancy scientists and managers made the hard decision, after less success with numerous alternative methods, to occasionally snare pigs. "Feral animals (mainly pigs and goats) are the most significant threat to the preservation of biological diversity in parks and preserves [in Hawaii]," noted Peter Vitousek, one of the nation's foremost authorities on invasive species. "Their feeding, trampling, and digging destroys native species, favors the spread of introduced species, and alters the dynamics of entire ecosystems."

As Vitousek's statement suggests, pigs and other alien animals can inflict pain and suffering of a profound sort, harming whole communities of animals and plants. Unfortunately pigs and their depredations aren't confined to Hawaii, nor is heartfelt opposition to the ways in which they're controlled limited to animal-rights groups, as the situation in California reveals.

On a cool March night, we walked through the night woods until we reached the vantage point Henry Coletto had selected before dark. From there we could remain hidden while watching the nearby intersection of three backcountry roads in the rugged hills about ten miles southeast of San Jose, California. Coletto spread an old horse blanket on the grass behind a big oak tree and he, Nicole Bowser, and I sat down to begin our stakeout.

A veteran officer with the Santa Clara County Sheriff's Office, Coletto was in charge of that night's operation. He spoke quietly into his walkie-talkie, checking on half a dozen law-enforcement people positioned in vehicles a short distance up each of the three roads. Bowser checked out the night-vision scope, which rendered a monocular image of oak woodlands bathed in eerie green light. Then we settled back and waited for the shooting to start.

Fortunately the bullets would cause no harm. The intended victims were wildlife decoys, including a fake wild pig that Coletto, a game warden, had arranged in a lifelike pose on the slope below us. Bowser and those law-enforcement officers worked for California Fish and Game, and their quarry that night was poachers.

As we waited, Coletto talked about flesh-and-blood wild pigs, the nonnative species that had drawn me, him, the other wildlife officers, and poachers to these northern California woods. These pigs run the gamut from feral animals that would look at home in a barnyard to the classic wild boars that inhabit hunters' dreams. No pigs evolved in North America; the peccary, native to the southwest, resembles a pig but belongs to a different family. Whatever their bloodlines, wild pigs cause all sorts of trouble.

"Most [big-game] poaching and trespass problems," said Coletto, "especially in the central-coast area, are related to pigs even more than deer." This is one of many reasons that most property owners dislike wild pigs, according to Coletto. On the other hand, some landowners make money by opening their properties to pig hunting for a fee. Others trade pigs for services. "There's a kind of bartering system," said Coletto. "Somebody will come out and fix a tractor and get a pig hunt in return." Still, the vast majority of landowners would rather pigs had never been brought to North America.

Several times our conversation was interrupted by the business at hand. The walkie-talkie would crackle and a disembodied voice would inform us that a vehicle was coming our way. We grew alert, but nothing criminal materialized.

Until a little after midnight.

A dark van cruised slowly through the intersection and came to a stop near the decoys. Coletto put his walkie-talkie to his mouth, ready to call in the troops as soon as the poachers took a shot. But after a few moments the van moved on, skulking up the road in front of us. After several minutes it returned. Again it paused in the intersection and then continued, this time turning up the road that ran about 100 feet to our right.

Then the rules were broken. Most poachers shoot pigs from their vehicles, but these interlopers pulled off about 50 yards behind us, got out of the van, and started sneaking back toward the decoys. I couldn't see them, but that doesn't mean they couldn't see us; Coletto had said that some poachers carry night-vision equipment, so I hugged the ground. Bowser was doing likewise right beside me. Suddenly her walkie-talkie sputtered to life, cracking the silence. She quickly rolled over, smothering the ill-timed transmission. Moments later we heard the crunching of dried leaves. The poachers were tramping into the woods toward the pig decoy. Watching them with the night-vision scope, Coletto had his walkie-talkie to his lips, waiting for the first shot. Several tense seconds elapsed. Suddenly Coletto blurted, "They're stealing our pig!" With a mixture of amusement and surprise, we sprang to our feet and watched drama turn to comedy: two people running toward a van, one carrying a wooden pig on his shoulder. Though these two guys had mischief rather than mayhem on their minds, Coletto had no choice but to summon the cavalry. In short order, two stunned young men were spread-eagled against a patrol car, probably thinking that they'd never tamper with a pig decoy again.

Though the stakeout story generally gets a laugh, many people are not amused that such effort goes into protecting pigs that are considered alien and invasive. There's a long-standing feud between anti-pig forces and the pig-promoting California Department of Fish and Game, which has overseen the management of pigs in the state since they were declared a game animal in 1957. People who are worried about pig damage think that protecting these animals makes about as much sense as protecting mosquitoes; critics of Fish and Game's pig management want pigs shot and trapped, free of legal restrictions. Of course, antipoaching operations such as this one protect public safety and the rights of landowners, yet many landowners and conservationists would like to see pigs eradicated. All cite the economic and environmental harm pigs have caused since the first specimens of *Sus scrofa* arrived in North America in the fifteenth century.

Christopher Columbus introduced swine to the New World on his second voyage, in 1493, when he left a small herd of pigs in the West Indies. Other explorers and the settlers who followed spread pigs throughout North America. In most places, owners let their pigs roam freely, a practice that continued into the twentieth century. When times changed and people penned up their pigs, many of these free-ranging hogs went unclaimed, forming the reproductive foundation for today's wild-pig population in America.

In 1912, an important addition occurred: hunters released European wild boar into the Hooper Bald shooting preserve in North Carolina. Many of these boars escaped. They and their descendants fanned out, interbreeding with established feral pigs to create the broad spectrum of strains that exists today. And since 1912, hunters have hastened the spread of European boar by releasing these prized trophy animals at many hunting sites around the country. Today more than 2 million wild pigs inhabit at least 23 states, with the greatest concentrations in the South, the Southeast, Hawaii, and California.

A veritable Noah's ark of other land-dwelling vertebrates has joined pigs in the United States. Many are so familiar that most people think they're native, such as house sparrows, some pigeon species, black rats; and others we know as pets, including some species of canary, chameleon, parrot, and gecko. Seeking new quarry, hunting enthusiasts and fish and game departments brought over ring-necked pheasants, Himalayan snowcock, Indian black buck, chukar, and Barbary sheep. Names of foreign animals, such as dogs, cats, cows, and horses are among the first words spoken by our children.

All of the above animals have established wild, self-sustaining populations in the United States. So have many others: 142 species, according to a study prepared by Stanley Temple and Dianne Carroll for the 1993 OTA report. That's 6.4 percent of the nation's total terrestrial vertebrate fauna (mammals, birds, reptiles, and amphibians), and it doesn't include the 50-plus indigenous U.S. species that have been transplanted to parts of the country where they're not native.

Of 142 species, Temple and Carroll studied the 130 for which the introductions were documented. The research revealed that 125 were brought into the United States intentionally and legally. Much of this was the innocent work of nineteenth-century animal enthusiasts, such as members of the Brooklyn Institute who appointed a committee to bring house sparrows over from Europe.

The history of the house sparrow's move to the United States is recounted in George Laycock's 1966 book, *The Alien Animals.* In the early 1850s, the Brooklyn Institute's committee imported a few birds from Europe. People lovingly cared for these birds, feeding and protecting them. Between 1852 and 1853, one member of the Brooklyn committee even kept sparrows in his house during the winter. Then their numbers began increasing when, in 1853, other bird lovers and immigrants who missed the familiar European sparrows brought over more of these nondescript songbirds. They landed in San Francisco, New York City, Galveston, Cleveland, Salt Lake City, and many other places. As the imports multiplied, people moved some of the sparrows' U.S.-born offspring to yet more sites. Soon this aggressive species began expanding on its own.

As populations of house sparrows grew dense, they began to cause problems. Attitudes toward them began to change. In the 1880s, a Massachusetts farmer reported that house sparrows munched every grain of wheat that he was growing for seed. One man who released the first house sparrows in his area later killed as many as he could in a futile effort to bring back native martins, swallows, and wrens, which the intruders had driven off. Other people came to despise house sparrows for the damage they did to gardens and decorative vegetation around buildings and houses. In 1887, the state of Michigan offered a penny for each dead house sparrow, no doubt sparking a run on slingshots and BB guns. People even tried to eat their way out of the problem, as house sparrows showed up on restaurant menus and market shelves. One New York newspaper reported that house sparrows "make excellent potpie." But evidently not nearly enough sparrows made it into potpies, because they flourished and spread throughout the nation. To this day they remain a serious pest.

Largely due to the twin plagues of house sparrows and starlings, Congress passed the Lacey Act in 1900 to control the importation of exotic wildlife species into the United States. But enforcement was lax and exceptions have been numerous until the last two or three decades, and many non-native animal species have entered the U.S. during the 20th century. Most so far have proven benign. Unfortunately that still leaves at least 38 species, including transplants, according to Temple and Carroll, that have caused problems. Not surprisingly, one of the animals high on their list of harmful species is the wild pig.

In California, wild pigs have followed the national pattern for swine introduction and spread. Spanish explorers and settlers brought pigs with them in the 1700s and let them range freely, establishing semi-wild herds. In 1924, hunters introduced the European boar in Carmel Valley. Subsequently, and often illegally, hunters have released European boar strains in many other parts of the state. Scientists estimate current California pig numbers at about 140,000 and growing.

Overall, despite a big decline during the long drought of the 1980s,

California pigs have been booming for the last 20 or 30 years. In the early 1960s, pigs inhabited fewer than 10 of California's 58 counties; today their grunts can be heard in at least 49. This pattern of initial slow growth followed by a sudden explosion typifies many invasions by non-native species.

Much of the success of pigs stems from their nature. One trait that allows pigs to prosper is their unmatched fecundity: they breed faster than any other large hooved mammal. Wild sows can breed when they're a mere six months old, kicking off a lifetime of producing litters of four to ten piglets up to twice a year. Such reproductive prowess often characterizes successful invaders.

Pigs also thrive because they are omnivores to a fault. It is no coincidence that our culture labels sloppy, voracious, indiscriminate eaters as "pigs." Wild pigs eat acorns, birds, earthworms, roots, carrion, cereal crops, lizards, mushrooms, tree seedlings, snakes, turtles, grass, eggs, frogs, herbs, salamanders, and sometimes their own piglets. On occasion pigs even prey on the young of other large mammals, such as cattle, sheep, and goats.

And this list of favored foods is far from complete. Just ask George Davis. For some 20 years, Davis has been the owner, manager, and chief bottle washer of Porter Creek Vineyards, a small family winery in Sonoma County, California. Davis is a pleasantly disheveled, middle-aged man who somewhat resembles Garrison Keillor in looks and manner. On the sunny spring morning when Davis showed me around his 35 acres, he was wearing an unbuttoned old flannel shirt over a worn T-shirt.

We puttered from his house up the hill to one of his vineyards in a spectacularly battered 1973 VW bug, his answer to the spiffy all-terrain vehicles used by most landowners these days. We stopped at the base of an oak-studded hill and stepped out into several acres of well-established Pinot grape vines: the site of the 1993 pig raid.

"The pigs were going around in marauding bands of as many as 50, hitting different places every night," said Davis. "Three nights before harvest they came in here and *boom*, it was blitzkrieg. They didn't bother the vines, but they like grapes, ripe grapes," he said, smiling ruefully between his black, bushy sideburns. "They got this entire vineyard. I only had enough grapes left to make half a barrel of wine." Laughing, he pointed at an adjacent vineyard. "Up there is my Chardonnay, and they hardly touched it. They just went over for a palate cleanser and then came back and finished off the Pinot." The pig raid cost him between $40,000 and $50,000 in lost wine sales—perhaps a quarter of his annual gross.

Davis had a fence around his vineyards, but pigs are notoriously adept at getting around fences—or even through them. Once Davis surprised four hogs in this same Pinot vineyard. They were rooting in the moist soil where Davis had been irrigating; later he found that they'd chewed up many of his plastic pipes. "When they saw me, they took off in a big rush. One made it out through the hole they'd come through, but the other three made their own holes. Ran right through the fence."

We walked over to look at one of the gaping exit wounds. The broken ends of the wire splayed outward toward the hills where the pigs took refuge. "Like a Sherman tank through chicken wire," said Davis. Recognizing the futility of repairing the fence, Davis built another one just inside the original. Made of thick, high-tensile wire stretched tightly between serious posts, this fence looked as if it would stop a charging rhino. "It cost plenty," said Davis, "but fencing works in direct proportion to the amount of money you put into it." Davis hasn't had a major pig event since putting in his great wall, and neighbors no longer refer to his place as "Porker" Creek Vineyards.

John Bucher had to resort to a fence, too. A fair-haired six-footer in his thirties, Bucher is a fourth-generation dairy farmer and past president of the Sonoma County Farm Bureau. Clad in the overalls and rubber boots that are de rigueur for a dairyman, Bucher climbed aboard his three-wheeler and drove us out onto his 350 acres. Jolting over a rough path, we ascended a hill verdant with spring grass. From the top, we saw the farm compound below us and the oak-speckled hills all around, but no pig damage. Then we reached the fence. Right on the other side pigs had rototilled a few hundred square feet of ground. The grass had been plowed under as the pigs had turned over the soil to a depth of perhaps six inches.

"That's just hors d'oeuvres," said Bucher. Gesturing at the pastures inside the fence, he said that a few years ago wild pigs had rooted up entire hillsides of his best grass. "Pigs go right for the good spots," said Bucher. "Where the cows would benefit most from the feed, the pigs are coming in and ripping it up."

He explained that dairy farmers in his region typically buy large amounts of feed because the terrain is too rugged to grow alfalfa or other feed crops. That makes the natural pastures vital. "We only have the pasture for a few months each year, so we really depend on it," said Bucher. "That's how we stay competitive, and if that's reduced by pigs, that's a huge economic loss." Hence the fence: "We spent about $15,000 on this fence," he said, "but we were losing [prior to building the fence] in the neighborhood of 10 to 15 thousand a year." Fence notwithstanding, the pigs still cost him 3 to 5 thousand dollars a year. And that doesn't count any long-term effects of soil erosion or the introduction of non-native species of plants and grasses that eventually could lessen the quality of the pasture. From year to year you don't notice anything dramatic, but it's a slow degradation."

About 150 miles south of Bucher's dairy, Wally Mark showed me yet another hefty pig fence. Mark is the director of the Swanton Pacific Ranch, a research and teaching facility in Santa Cruz County that's part of California Polytechnic State University. The fence was one of three under construction, a joint effort of the ranch and the organic farmers who lease the fields. During the previous season, organic squash, sweet corn, cabbage, rosemary, basil, and other high-value crops had been trashed by pigs, with an assist from deer. The farmers said they either would have to build fences or abandon the farm. But the fences won't end their pig problem. Every year the wild pigs will cost them because the new fences forced the farmers to relocate some of their equipment roads, which "means they'll lose two or three acres of productivity," according to Mark. "That's about 10 percent of their land, which is quite significant given their profit margins. And that's year after year after year."

Pigs have plagued the Swanton Ranch for years. Mark recounted the time hogs ate the hearts out of $45,000-worth of romaine lettuce. He pointed to the abandoned artichoke fields on a nearby hill where, in the winter of 1995 to 1996, pigs destroyed seven acres of drip irrigation pipes. The ranch gave up on growing artichokes there. Mark said that the 150 to 200 pigs on Swanton Pacific's land eat so much grass that the ranch has reduced the number of cattle they run on their 1,900 acres of range. They've even lost cows to pigs: "If there's a breach birth or something," said Mark, "and the cow is down struggling with it, the pigs will come up and eat both the calf and the cow."

Given Mark's litany of loss, one would expect to see pig fences all over the ranch, but the three he mentioned are the only ones. Fences cost too much to use for the protection of large areas or less valuable crops. But Mark has an idea for a much less expensive way to keep pigs out.

Sheep dogs guard sheep against coyotes; Mark hopes that crop dogs will guard crops against pigs. In other places, crop dogs have successfully discouraged deer from browsing in orchards. Mark pointed to the middle of a big field where an Akita and a German shorthair–dalmatian mix were pulling sentry duty. Mark decided on the use of two dogs after seeing a pattern in the encounters between a friend's golden retriever and wild pigs. "The dog would chase after a boar and they'd disappear into the woods, and then pretty soon the dog would come running back with the boar chasing him." The presence of more than one dog prevented this embarrassing turnabout.

After being trained to stay in the field, the dogs lived there day and night during 1996. "We had sweet corn in here last year and had no problems," said Mark, "while a grower up the road lost all of his crop to pigs." Though he's not entirely sold on crop dogs yet, Mark plans to try more dogs in more fields soon.

Usually wild pigs stick to rural lands, but at times they venture into suburbs and towns. Often they are drawn by moist soil, and during the dry season, freshly watered grass is a pig magnet. A few years back, the residents of a retirement community near Santa Rosa were up in arms when pigs plowed the subdivision's tidy lawns. The pigs discovered the golf course, too. "It looked like bombs had been going off," said one witness. Down in Marin County, a lone pig repeatedly savaged the fairways at the Meadow Club. After each attack, it took four people four hours to fix the mess.

While many homeowners have skirmished with pigs, Bill and Shirley Walker had to wage war. From 1993 to 1996 pigs assaulted their 20-acre hilltop estate in the coastal town of Aptos. "We were being overrun," says Bill. "They were tearing up the lawns, the flower beds, everything." Recounting numerous surprise encounters with pigs, Shirley added, "It got to the point that I was very nervous about going outside."

With mixed results, the Walkers took matters into their own hands after getting the proper permits. A live trap they set worked fine, but it caught their German shepherd twice. Bill shot a few pigs, including one from their garage and one on the front walkway, but he also experienced some technical difficulties. One night he tied a flashlight to the end of his rifle to illuminate his quarry. Unfortunately the light was too near the end; when he fired he blew off the top of his flashlight.

Bill's first kill was perhaps his most memorable. "We were up in our bedroom when we heard all this ruckus and grunting early one morning." He looked out and saw four pigs rooting around beneath the second-floor window. Grabbing a 9mm pistol, he shot right through the screen, hitting one of the pigs. Clad only in his skivvies, Bill ran into the yard in pursuit of the wounded beast. As he stalked through the bushes the big sow suddenly charged at him. He sidestepped it, fired twice as it thundered past, and watched it disappear into another part of the garden. After a long search, Bill finally spotted the sow standing next to a low retaining wall behind some bushes. He prepared to shoot again, but something about the pig's posture seemed odd. Bill crept closer and closer. When he got within a few feet of his quarry, he lowered his pistol. The sow was dead. Apparently it had propped itself against the wall before expiring.

The Walkers began to win the war when they hired Jim Riveland, a locally renowned pig hunter. On and off for more than two years, Riveland worked the estate. Some nights he lay in wait in the back of his truck or in a tree blind. But the hunting was tough. There were safety concerns in an area dotted with houses; and legal restrictions limited shooting pigs to only when they were on the Walker's acreage. Still, Riveland killed dozens of pigs; counting those shot by Bill and a few others, 47 fine specimens of *Sus scrofa* were taken on the estate. Now the Walker's lawns and gardens are recovering, and the pigs are mostly gone.

When pigs carve up lawns or gobble up grapes, the loss is mainly financial, with a dash of emotional suffering. But pigs harm the environment as well. At least, many scientists, conservationists, and land managers think so—though some officials at the California Department of Fish and Game disagree. "I have yet to be convinced that pigs have a negative effect on the environment," said Jim Swanson, a senior wildlife biologist for Fish and Game. He added the caveats, however, that an overpopulation of pigs can create problems and that the rooting of steep slopes may cause erosion.

Wally Mark would agree with the remark about steep slopes, only he'd delete the note of uncertainty. Standing beside a feeder stream above Scott Creek on Swanton Ranch, Mark nodded to a section of the stream that had been churned into a mud pit. "That's a pig wallow," he said. "We've got some areas up on the rangeland where pigs have the whole thing chewed up into a big muck hole; so when it rains, all the water flows across that, right into Scott Creek."

Farther along the creek, Mark pointed to a landslide that had occurred the previous winter. A bare, brown stripe ran for hundreds of feet down the steep hillside toward the creek. Pigs had severely rooted the slope at the top of the slide. Mark thinks the animal-plowed ground allowed the rain to catch and saturate the soil instead of staying on the surface and running off, though he said he can't be sure without doing research. The soils of the Santa Cruz Mountains are slide prone, and he thinks pigs can, so to speak, push them over the edge. Erosion above Scott Creek poses a particular danger because this shallow, shaded coastal creek shelters coho salmon, which have been placed on the federal threatened-species list. Sediment can cause a number of problems for coho, notably smothering the eggs, which are laid in gravel beds.

Other landowners and managers report similar experiences with sediment-laden runoff due to pig rooting and wallowing. George Davis pointed out a couple of pig-rooted hillsides around his vineyard that had eroded badly during the rainy season. He also pointed out a double standard. "In steep areas where you'd never allow horses because of the damage they do to the land, pigs come in and do more damage." Greg Giusti, an ecologist for the University of California cooperative extension office in Mendocino County, notes another double standard. Gesturing toward a slope pigs had plowed, he said, "The law says a guy can't come out and do this with his tractor. Of course pigs are smart, but they aren't good readers and don't know what the law says."

Rooting on level ground also draws protests from many conservation-minded citizens, which particularly vexes Jim Swanson and others at Fish and Game. Swanson asserts that even though torn-up pasture may look damaged, it quickly recovers with no lasting harm. Phyllis Faber, a biologist who is vice president of the California Native Plant Society and a fierce critic of Fish and Game's pig policies, disagrees. The worst damage occurs when native plants inhabit the land under assault. Faber cites many cases on the lands around her home in Marin County, where pigs root up areas of calypso orchids and other uncommon native plants. In some instances the native plant community will recover. In other cases, especially if the pigs return again and again, the natives probably won't make it. Even if the plants survive for the time being, the rooting adds stress to communities that are already highly vulnerable. "Pigs," said Faber, "are one more factor in the decline of a world-famous and unique flora here in California."

Furthermore, the plants that sprout on the bare soil of a rooted site may not be the same plants that grew there before the pigs. "The minute pigs plow up the ground," she added, "non-native weeds move in. Yet another area gets converted from a native flora to a weedy flora."

Back on the Swanton Ranch, Wally Mark and colleagues constantly battle this unwanted conversion. Their counterattack is to sow seeds of native perennial grasses on rooted sites as they roam their land—but the ranch contains 1,900 acres of range. "We miss spots, and we get places that were rooted and not seeded with natives," said Mark. Typically exotic thistles move in. If the infestations are small, savvy grazing management usually can eradicate them. For example, Mark explained that sometimes he'll put 200 head of cattle onto the infected site, fence them in for a day or two, and let them tromp it down. Then he'll reseed. Other times the invasions get out of hand: "They'll take over. We've had several places just turn into thistle thickets that the cattle wouldn't even go into." Native seeds are expensive, but so are other options, such as mowing, digging, and burning. And they all take time and labor. "Dealing with thistles is not an inexpensive operation," said Mark. "And we're taking care of something that we wouldn't have to do if pigs didn't root it."

Little research has been done in California on the effects of rooting on plant communities, but a fair number of studies have been conducted in Great Smoky Mountains National Park, where land managers have been wrestling with pigs since the 1940s. "It's like virtual clearcuts of the understory," said Peter White, professor of biology at the University of North Carolina, Chapel Hill, who has studied pigs and plants in the park. White said that in the spring and early summer, pigs congregate in the high-elevation hardwood forests, so their rooting is concentrated. In these areas, says White, "The amount of plant material, the number of living plants, the leaf area of plants, and the diversity of plants are all decreased by pigs." In pig-free gray beech groves, vegetation covers 80 to 100 percent of the ground; after pigs move in, that ground cover drops from 2 to 15 percent. Echoing Faber, White worries about harm to uncommon native plants such as Clingman's hedgenettle. A rare species once found only in the park and a few adjacent counties, it no longer grows in areas rooted by pigs.

White has also noticed subtle changes in the composition of plant communities, which are due to the differences in the ability of plant species to recover from rooting and the pigs' taste for certain plants—they churn the leaf litter into a big mass, which favors particular kinds of seeds and soil organisms. White also suspects that pigs might affect nutrient cycling. For example, scientists think that rooting may lead to higher rates of decomposition and the release of nitrogen from the rototilled soil. They speculate that nitrogen produced this way may leach into groundwater, decreasing soil fertility. One wonders how long the list of concerns would be if more research were done in more places.

Pigs harm native animals as well. Sometimes the damage is direct. Remember, frogs, snakes, lizards, deer, turtles, and other animals show up on pig menus. A study in Great Smoky Mountains National Park showed that pigs reduced the density and diversity of salamanders; some researchers worry about pigs munching the eggs of ground-nesting birds, though studies have been inconclusive. Occasionally the effects of hogs' dining habits severely change a local indigenous population: one study showed that 80 percent of sea turtle nests on Ossabaw Island, off the Georgia coast, were devoured by pigs. The evidence of this kind of predation is spotty, but researchers feel that there is potential for significant damage.

Harm definitely ensues from pigs' gluttonous consumption of acorns. Pigs crave acorns the way kids crave ice cream. Assuming that oaks produce a typical crop, pigs pretty much live on acorns during fall and winter. Unfortunately so do deer, black bear, squirrels, and other wildlife. Though no research has determined precisely how this competition plays out, scientists are convinced that it sometimes harms wildlife. Reg Barrett, a biologist and wild-pig authority at the University of California, Berkeley, points out that acorns can be vital to deer nutrition. Acorns provide the high-energy diet deer need during their breeding and gestation periods. In years when there aren't enough acorns to go around, deer may suffer due to competition from pigs. Given that California's deer population is declining, such additional pressure is unwelcome.

Charismatic megafauna, such as deer and bears, are not the only animals adversely affected by pigs. Peter White notes that many sorts of insects, soil dwellers, small mammals, and other unassuming creatures depend on existing plant communities. When rooting alters the nature of these communities, the lives of these uncharismatic minifauna probably are altered, too. Looking at all the environmental impacts of pigs, it becomes clear that these European imports can fundamentally change ecosystems.

"Fish and Game would like to see more pigs on public lands, except where they cause undue problems." These words came from Brian Hunter, the fittingly named regional manager of the central-coast region for the California Department of Fish and Game. That makes him the person in charge of the counties where most of California's wild pigs live. Given all the problems pigs do cause, Hunter's statement worries many wildlife biologists, ranchers, conservationists, farmers, rural homeowners, and land managers—especially when they also consider that Hunter agrees with Jim Swanson that pigs seldom cause "undue" problems.

That Hunter specified public lands wouldn't soothe many private landowners. Public and private properties are mixed throughout California, and pigs don't respect boundaries. Besides, Fish and Game clearly doesn't mind the presence of pigs on private lands: the department's main goal regarding pigs is to persuade landowners to open their properties to sport hunters.

I recently visited Fish and Game's Napa Valley office, headquarters for the central-coast region. Mounted deer, grizzlies, ducks, elk, and even a polar bear inhabit the halls and offices. Though the department also is responsible for all state nongame wildlife and resources, hunting and fishing come first, as they do in most state fish-and-game agencies. Many of these agencies get low marks when it comes to exotics as well. Bob Ferris, director of species conservation for Defenders of Wildlife, recalls a presentation made by California Fish and Game at a small meeting of wildlife professionals. The department had hired a public-relations firm to work up an ad campaign to make the public more aware of wildlife. A member of the firm presented the campaign's slide show. It featured striped bass, pheasants, and a couple of other wildlife species that would serve as the focus of the campaign. When the lights came on, the smiling PR guy proudly turned to the audience and asked, "Well, what do you think?" The crowd was dead silent, reports Ferris. Finally somebody spoke up and pointed out that all those "wildlife" species being promoted were non-native—not to mention that they were all game animals.

After waiting briefly beneath the glassy stares of the four bighorn sheep trophies in the reception area, I was ushered to Brian Hunter's office. Hunter is a tall, mustachioed, white-haired man who has worked for the department for decades. We were joined by Jim Swanson. Hunter spoke of the joys of pig hunting, which he has enjoyed since he was a boy. "It's fun. It's a lot of fun. And it's exciting," he said with a broad smile. "You can do it at a time of year when you can't do anything else, since it's after duck season and before the fishing gets good." He talked on about the romance of the baying hounds, the chase through the woods, and the intelligence and occa-sional fierceness that make pigs a worthy quarry.

Hunter was more concise about Fish and Game's pig management policy. "We have no real official policy," he said. The department has been working on a management plan, however, and hopes to produce it soon. Despite the lack of an official policy, Hunter and Swanson made it plain that Fish and Game promotes sport hunting as a key to pig management.

Pigs enjoy a lofty status in the California hunting community. Sport hunters kill from 30,000 to 40,000 annually, making pigs a close second to deer—and some observers think pigs would be number one if illegal shootings were included. Still, the demand for pig hunting far exceeds the limited opportunities due to land-ownership patterns. Most pigs inhabit the oak woodlands. Ranchers and other landowners long ago claimed a large majority of this habitat, and in recent times many people have moved into these attractive rolling hills. Today about 90 percent of the wild pigs in California roam private land.

Fish and Game would love to see property owners open their land to licensed hunters. Rock Springs Ranch epitomizes one model that has been tried by a few large properties around the state. This 19,000-acre spread encompasses a rugged hunk of the Diablo Range in San Benito County, California, about 75 miles south of San Jose. Ken Range manages the ranch's hunting program, which was our topic of conversation one warm spring day as his vehicle conveyed us along a rough track in the remote hills.

"The owner was looking for a way to make the ranch pay," Range explained, "given California's high property values. The cattle business has such a thin profit margin; there's no future in cattle unless you inherited land and own the cattle free and clear." So in 1993, Range set up a major hunting program, built mostly around game birds. The ranch employs nine full-time employees for hunting. Guests stay at a swanky lodge, feast on fare fixed by a chef who worked with Wolfgang Puck, hunt with guides, practice shooting at a range, and have their kills prepared for transport in an elaborate game processing shed. In 1997, Rock Springs Ranch got out of the cattle business entirely.

Pigs account for only 10 to 15 percent of the hunting, but they're a major source of income. "There's very little management for pigs," said Range. "A bit of water in summer, and we don't run them out of the barley fields. They're very profitable." A pig weekend runs more than $600 per person, and Rock Springs books about a hundred weekends a year. Range enjoys doing that arithmetic. He added, "There's such a demand for hog hunting that there's no need to advertise it. I have to turn hog hunters down every day."

Range's last statement is the kind of remark that brings Brian Hunter to the edge of his chair. He hates hearing about would-be pig hunters being turned away. He knows that Rock Springs Ranch and its ilk can meet the needs of only a small segment of the hunting public. Swanson and Hunter from Fish and Game want to match up average-Joe pig hunters with landowners for whom hunting would be a sideline, not their main business. Fish and Game sees it as a win-win situation that would help satisfy the high demand for pig hunting, expand the reach of their management program, and control unwanted wild pigs.

Sounds good. Some landowners, even critics of Fish and Game, have tried this approach. John Bucher, the Sonoma County dairyman and a pig hunter himself, gave it a shot several years ago but quickly ended the experiment. "I just can't manage this by allowing 20 different sport hunters back here, especially people I don't know. And they all want to come out on a Saturday or Sunday, then sometimes no one comes out. I've got a business to run. I can't sit there and be trying to do an appointment book for a bunch of pig hunters. It was a logistical nightmare. And some of these folks . . . well, I shouldn't have allowed them back here. They were not responsible."

Ecologist Greg Giusti expands on landowners' concerns over the minority of hunters who are careless and cause trouble. He cites fires started by hunters, livestock stampedes begun by hunters' shots, lost dogs roaming around private property, damage to water troughs, and fence gates left open. "You get some real lowlifes, too," Giusti added. "They're out there at two in the morning drinking beer, cutting fences, and chasing hogs. They're not the kind of people I'd invite to my house for Christmas."

Henry Coletto, the Santa Clara County game warden, tells about a small subculture of pig hunters who assert their toughness by killing big hogs without using guns. Late one cold December night, Coletto and a state game warden had lain in wait on a back road in the Gilroy area of California. "We saw the fence was cut," said Coletto. "We stayed till about four o'clock, then the poachers came out— and they had one of the biggest pigs I'd ever seen. They had a winch on the front of their truck, and they winched the pig up into a tree and then drove the truck underneath. We learned later that the pig weighed 403 pounds.

"One guy came out to the road to see if anybody was there. The guy jumps over the fence, and he walks up the road right toward me. But it was pitch dark, and he probably wasn't but 30 or 40 yards from me when he saw my vehicle. He took off running and jumped back over the fence. I threw my headlights on and took off chasing him in my vehicle. I drove right through the fence." (The landowner had told Coletto to do whatever it took to catch these longtime poachers.)

Coletto laughed as he continued. "The barbed wire got tangled up in my wheels. It was like a rubber band, all that wire tangled around my front end; and this great big four-by-four [fence] post shot out like a spear and went flying right past this guy. Missed him by about three feet. When that happened he stopped running. Then we went in and secured the situation.

"Both of the fellas were in their late 20s, maybe early 30s," Coletto continued. "This one fella had a kind of dagger, about a foot and a half long and double-edged. The pig had 18 holes in its side. I said 'How'd you stab it that many times?' He said, 'I rode it.'"

Poachers often hunt with four to six dogs, some to locate and run down the pig and some to catch it. These two men even had leather vests for their dogs, to protect them from slashing tusks. "Their catch dog, a pit bull, was on the front of the pig's head," explained Coletto, "and these guys have so much confidence in their dogs, they'll get on. So he was saddled on that pig and he stabbed it in the lungs."

Coletto found the 23-inch tail—evidence of the boar's size—on the dashboard of the poachers' truck. Even though the owner of the tail was under arrest, he begged Coletto to let him keep it so he could swagger into his favorite bar and brag about the monster boar he'd slain.

Hunter and Swanson acknowledge the landowners' concerns about opening their properties to hunters. "They feel it's like letting a stranger in your backyard," says Swanson. That's why Fish and Game promotes the use of licensed guides for property owners who otherwise aren't comfortable opening their land to hunting. Hunter says guides can assume the burden of organizing hunters and overseeing their behavior. Peter Bradford follows this model. He leases pig-hunting rights on his family's ranch in Mendocino County to a guide for $6,500 a year. A poster boy for Fish and Game's approach? To an extent. But Bradford paradoxically is also a vocal critic of Fish and Game's pig regulations. "Fish and Game is a bureaucracy out of control," said Bradford one day as he showed me a pig-rooted pasture on his land. "They're a juggernaut. It's going to take a lot to turn them."

Bradford appreciates that $6,500, but he'd much rather not have pigs on his land at all, and the sport-hunting operation hasn't come close to curing his pig problems. Other landowners who allow pig hunting on their property have also found it only marginally effective in controlling pigs. Wally Mark reports that hunters shoot 60 to 70 pigs on the Swanton Ranch every year, but he explained that this just allows more piglets to survive because the killing of 60 to 70 adults frees up a lot of resources.

John Bucher, ex-president of the Sonoma County Farm Bureau, expresses the prevailing view: "Fish and Game should allow farmers to manage pigs as if they were a pest. If they're doing damage, we should be able to get rid of them and be done with it. No tags, not even a hunting license. Like going out and buying poison and getting rid of a rat." Bucher, Bradford, Mark, and just about every other landowner and land manager in pig country want Fish and Game to loosen their regulations governing the shooting and trapping of pigs. In recent years, these regulations have been relaxed, but not enough to satisfy those who battle pigs year in and year out.

Even quiet-spoken George Davis, the Sonoma County vintner, reproached Fish and Game: "We're up against this bureaucracy of Fish and Game," said Davis. "They don't seem that interested in helping us." Davis wishes Fish and Game would get its priorities right. "I think the primary consideration should be the land itself. This is our nonrenewable resource. If it's a question of more pigs versus more deer or more wildcats, I think the balance should be in favor of the native species, which evolved on this land and are part of it and not so antagonistic to it. . . . [Fish and Game] seem much more interested in having their pig-hunting program."

Other people who also criticize California's pig policy think that money motivates Fish and Game. "They're absolutely obsessed with tag money," said Dave Chipping, a university geology professor and an officer of the California Exotic Pest Plant Council. Pig tags are essentially permits that hunters buy, in addition to hunting licenses, to hunt pigs legally. Peter Bradford, the Mendocino County rancher, expanded on Chipping's succinct evaluation. "The decision to call pigs a game animal is a monetary one. Deer hunting is declining . . . and Fish and Game sees pig hunting as a huge moneymaker."

Brian Hunter, Jim Swanson, and others at Fish and Game deny that they see dollar signs when they look at a snorting boar. They point out that pig tags raise only about $300,000 a year for a $190 million budget. The department's adversaries counter that every $300,000 counts when one considers California's current funding situation: Fish and Game gets fewer tax dollars than in the past and must generate more of the money for its budget. Furthermore, Fish and Game may see potential for much bigger returns from pigs. As Greg Giusti said, "There's no doubt pigs are a marketable commodity and a highly prized one." If Fish and Game's goal of increasing the number of pigs on public lands and the number of pig hunters on private lands comes to pass, *Sus scrofa* could produce millions of dollars.

However prominent the role of money, something deeper also plays a part in Fish and Game's passion for pigs and in their reluctance to cede more control to landowners. If landowners are given more control, there will be less opportunity for sport hunters to shoot pigs, and less hunting in general. That concerns many hunters, who already believe that the sport is fading from contemporary American society. Fish and Game's Brian Hunter is among those concerned sportsmen. "We're interested in seeing hunters happy, because we're worried about the future of the hunting heritage. If hunters aren't happy, they don't hunt—and that means they're not going to train their kids to hunt and that means that the people who are against hunting are likely to be more successful."

Jim Swanson, Brian Hunter's colleague, counters criticism of Fish and Game's management policies and regulations with his own accusations. "[Landowners] want the government to come and solve the problem for them. It seems like they want to complain more than to solve the problem." Hunter agreed on both counts. He called landowners a bunch of whiners: "They keep demanding [less restrictive regulations], but if we gave it to them, they'd still be wanting someone [from the government] to come out and kill their pigs."

As problems with exotic animals have increased, disputes over the management of non-indigenous species have prompted sign waving and table pounding across the country, and within and between organizations, governmental agencies, and landowners. Battles over the fate of the mountain goats in Olympic National Park have been fought for years; cat lovers all over the U.S. are hissing and scratching over the killing of feral felines; and back East, people have been protesting the removal of mute swans by the U.S. Fish and Wildlife Service and state wildlife agencies. Once when University of California ecologist Greg Giusti was roaming a remote corner of California, he got to talking with an old cowboy "who had calluses on his calluses and a Stetson with lots of miles on it." Giusti and the old man hit on the subject of wild horses and burros and animal activists' fervent protection of them. "In these parts," said the cowboy, "you can shoot a man, but don't shoot a wild horse or you'll get in real trouble."

Animal activists are passionate about non-native animal management. It was passion that in 1994 impelled Alex Pacheco of PETA to storm the Nature Conservancy's annual meeting, and it was passion that pushed PETA protestors to throw ersatz blood at the TNC headquarters. "The issue above and beyond anything else is that the animals suffer so severely," said Tracy Reiman, a campaign manager for PETA and organizer of the infamous 1993-1994 campaign against the TNC. "They were left for so long to suffer and die."

The National Park Service was the first to use snares; they pioneered their use in an effort to control non-native pigs and goats in national parks on the Big Island of Hawaii. The snares were designed to tighten around an animal's throat and kill it in a few minutes. U.S. Fish and Wildlife Service, the Hawaiian Division of Forestry and Wildlife, some private landowners, and the Conservancy have employed the same snares. Pig killing of any sort bothered PETA; and imagining snared pigs thrashing in pain before dying was even more disturbing. What ignited PETA finally, and sparked the anti-Conservancy campaign, were the occasional snares that didn't work properly, prolonging the trapped animal's horrible death. Rieman claims that the snares never kill pigs as rapidly as they're supposed to.

This suffering distressed TNC staffers, too. "We made the very difficult decision to use snaring as a last resort," said Maria Naehu, director of communications for the Conservancy in Hawaii during the PETA campaign. "For years and years we relied on fencing and hunting at our preserves on Maui and Molokai, where pigs and goats are a huge problem. But we were clearly losing the battle. The damage was going up and up, and pig numbers were skyrocketing."

Completely fencing the preserves was impractical due to their size and topography; extremely rugged mountains and dense tropical forest cover much of the TNC properties. The topography made hunting less successful, too. And the wild pigs in Hawaii, like their kin in California, are wily and exceptionally fecund. Extensive hunting, including game drives and helicopter hunting, would reduce the population—but remaining pigs were cautious and would scatter. Hunters would often return empty-handed. In short order, the pigs would reproduce their way back to previous population levels, levels at which 60 percent of the forest showed signs of recent pig activity. Then the Conservancy once again would initiate intensive hunting, heading for another spin around the vicious circle. "What we'd created," said Naehu, "was, in essence, a perpetual-killing program."

Feeling trapped themselves, the Nature Conservancy began the snaring program in 1989. Using helicopters to reach remote areas, they saturated the least accessible portions of their preserves with hundreds of snares. In about two years, they killed several hundred pigs and goats, lowering populations to the point that they could be controlled with a little maintenance hunting and trapping. "That was our theory," said Naehu, "that if we got the numbers very low, over time we'd have to kill far fewer pigs—and that's what happened." Now only about 50 pigs a year get snared and less than 10 percent of Conservancy land shows signs of recent pig activity.

Snares work well for many reasons. For one thing, they're on duty 24 hours a day, 365 days a year. For another thing, pigs don't get wise to them as they do to hunters or cage traps. Perhaps most important, snares can be left on guard in areas that can't be visited frequently. This was part of the reason TNC staffers only checked the snares every three to six months, a practice that drew fire from PETA. PETA insisted that if the Conservancy continued to use snares, at least they should check them daily to shorten the misery of any improperly caught pig or goat.

Even if the terrain had been conducive to daily checks, TNC would have resisted them. U.S. Park Service research indicates that repeated checking somehow causes pigs to avoid snares. More to the point, in light of the Conservancy's overall mission, frequent visits damage Hawaii's unusually delicate ecosystem. For millions of years, nothing heavier than 25 pounds walked through Hawaiian rain forests. Native vegetation did not evolve to withstand 160 pounds of pressure wearing hiking boots, which might also carry alien plant seeds into these vulnerable rain forests. Besides, if people could actually get back into that countryside every day, TNC would have used hunters instead of snares. The Conservancy's policy is to use snares only where nothing else will work.

PETA accused TNC of violating this policy. "Plenty of humane alternatives to snaring existed," said Reiman. "But these alternatives cost more money, and snares cost very little." The unwillingness of the Conservancy to spend more money in order to alleviate the suffering of pigs became a major theme in the PETA campaign. "The Nature Conservancy has tons of money," said Reiman. "It is their duty to use that money to remove these animals in the most humane way possible, even if it is expensive. It's not going to make them go broke."

"Any time we'd tell PETA that something was impractical," counters TNC's Naehu, "they'd say, 'You mean it's hard and expensive and you just don't want to do the hard thing.' It's very frustrating talking to these people, because there will be an absolute, black-and-white fact sitting there and they completely ignore it."

Naehu cites the example of birth control. Contraceptive vaccines have been used successfully to combat overpopulations of deer, wild horses, and other wildlife. PETA considered this a viable alternative to the snaring of pigs on Hawaii. TNC staff pointed out that no contraceptive vaccine for pigs existed. Even if vaccines had been available, they wouldn't have worked in this situation.

Contraception will not bring pig numbers down to zero or near zero. Pigs are notoriously promiscuous; and with weak bonding between mates, all males and females would have to be inoculated. That's virtually impossible. In addition, each pig would have to be regularly reinjected. Even if this was possible, at best the result would be a tapering off of an otherwise vigorous population, not a reduction to zero or near zero. "Besides," said Naehu, "if we could get close enough to inject the pig, we'd shoot the pig." She added, "PETA would completely gloss over [the facts about contraception] and continue to argue, continue to write in their literature that there are vaccines and the Nature Conservancy is choosing not to use them. They used lots of misleading and incomplete information."

As another alternative, PETA offered to assemble an army of volunteers to form a human chain and drive all the pigs and goats off Conservancy lands. The notion is clearly unworkable. Multitudes of people would have to be ferried by helicopters to the mountaintops. From there they would need to walk together down treacherously steep slopes through a tropical tangle of vegetation. Even if such an impossibility could happen, what would it accomplish? The pigs and goats would be driven to the land of some neighbor who didn't want them either. And what would stop them from returning to the TNC preserves? When the Conservancy declined to employ the human-chain approach,

PETA again accused TNC of rejecting viable alternatives to snaring.

In an effort to sort through the rhetoric surrounding humane alternatives, the Conservancy formed a consortium. Among its members were the National Park Service, local hunters, the Humane Society of the U.S., veterinarians, the U.S. Fish and Wildlife Service, the Hawaiian Humane Society, the Conservancy, the state division of Wildlife and Forestry, and PETA. By consensus they selected a consultant, and after a meticulous review of control technologies, the consultant and consortium found no replacement for snaring in remote Hawaiian rain forests. PETA quickly labeled the consortium a sham. "We went in good faith," said Reiman. "However, it did not appear that the Nature Conservancy was really interested. They were using the consortium as a public relations move."

Another of PETA's claims challenged the basis of the Conservancy's anti-pig efforts. Reiman summarized that claim: "We feel that though pigs are not native, they are not doing any more damage than snaring them does." She elaborated, pointing out that when pigs are snared they thrash about, tearing up the soil and vegetation. She equated this damage to that done by the pig population in general. A snared pig, however, could rip up only a few square yards of forest floor. All the pigs ever snared by the Conservancy might have wrecked a couple of thousand square yards—an area that a free-roaming herd of pigs could demolish in a day or two.

"In a lot of instances [animal welfare] is almost a religion replacement," Bob Ferris of Defenders of Wildlife said. "People clamp on to the dogma and nothing can sway them." Ferris and Defenders have had numerous run-ins with animal-rights activists over wildlife issues. "You go home at the end of the day thinking you've been doing really good things," he added, "but you almost feel bad about some of the stuff you're doing because you've been pounded by these people. They can be very self-righteous."

Holt, Naehu, Ferris, and others who have dueled animal-rights groups are quick to point out that significant differences exist among the various organizations. For example, the Humane Society of the

U.S. (HSUS) opposes snaring, but they understood TNC's dilemma and didn't go after the Conservancy as aggressively as PETA did. Instead HSUS has worked with TNC to find more humane alternatives. The American Society for the Prevention of Cruelty to Animals is another powerful animal-welfare organization that appreciates the dilemma posed when non-native animals must be killed to save native wildlife and ecosystems. "If I told you I had an answer [to that dilemma], I'd be lying," said Stephen Zawistowski, a senior vice president at the ASPCA.

Though Alan Holt took his lumps from PETA during their campaign, he appreciates their concern for animals, even non-natives, even the pigs that are harming the rain forests he loves. "PETA makes an important point," said Holt, "that all of us in the business of managing to minimize the impact of alien species should keep in mind: These are not evil animals; they don't have bad intentions. Their impacts are the result of human actions in bringing them here or allowing them to come here. Vilifying them is at best a motivational strategy for a bunch of humans, just like talking about the enemy during a war and making them evil in order to kill them.

"On the other hand," continued Holt, "people who advocate the rights of animals need to broaden their view beyond animals with warm blood and hair and eyelashes to all animals. And to consider the rights of the native ecosystem. They might even extend their consciousness to plants.

"They say they're here to talk about suffering and cruelty, and we say, 'So are we.' We're weighing the total suffering that's happening out on the landscape and choosing to kill these pigs as the least-suffering alternative. And if you want to compare suffering, let's go look at the forest. Go look at the birds that live there that can't find larvae in the understory to feed their chicks because the pigs have removed the understory, or birds that are dying from avian pox [spread in part by pigs] . . . their toes fall off and their beaks rot away."

Interestingly, in the middle of lambasting the Conservancy, PETA's Reiman did acknowledge, momentarily, TNC's noble goals:

"The Nature Conservancy is very concerned about precious, endangered species, whether plant or animal. But an animal that is non-native, or not that cute, or which people don't care as much about, like wild boars, the Conservancy doesn't care about at all. They'll use the most barbaric method possible to kill them. But we have to be concerned about all of the animals, all of the species out there, regardless of what they look like or how people see them."

The resemblance in language and thought that exists even between the Conservancy and an animal-rights group as extreme as PETA reveals an essential kinship between the two types of organizations. They appear to part company when animal lovers, confronted by a creature's pain, become blind to ecological complexities—and to the future. When discussing TNC's point that the lushness of the Hawaiian rain forests makes access difficult, Reiman said, "The Conservancy claims that the animals have to be removed because they're destroying everything, yet in the next breath they're talking about how the woods are so thick they can't even get in there." Equating lushness with ecological health reveals a poorly informed view of nature. On Hawaii, much of that lovely greenery is alien, an environmental curse. Reiman's point also neglects the problems that are hard to see. For example, scientists found that tree-fern trunks, hollowed out by pigs feeding on their starchy cores, become water troughs in which mosquitoes breed—mosquitoes that carry avian malaria, which is devastating Hawaii's bird populations. "Some people who really love wildlife but don't know much about it do things that actually damage wild species, out of ignorance," said Bob Ferris of Defenders. Certainly a comprehensive education on the intricate workings of nature will bring these natural allies closer together.

PETA eased back on the throttle in 1995, after a couple of years of fiery campaigning against TNC's snaring program. In the end PETA's efforts mattered little. More than anything the campaign diverted Conservancy resources. "It took a lot of work on our part," said Naehu. "Where PETA hit us the hardest was in staff time. Staff time and misery." She spent a huge chunk of her time on the issue, including countless hours on the phone explaining TNC's position to angry Conservancy members who had received inflammatory literature from PETA. Most of them soon grasped the complexities of the situation and very few quit the Conservancy. And most of the Conservancy staff rallied around their beleaguered Hawaii colleagues. PETA's pressure on big donors and on TNC chapters outside Hawaii likewise fell flat. No major corporate sponsors or foundations pulled their support.

Campaigns protesting the control of harmful exotic animals don't always fail. On the contrary, pressure from animal-rights groups strongly influences the management of many alien species. The Park Service long ago would have removed the few hundred mountain goats in Olympic National Park if it weren't for the objections of animal-welfare groups. Federal law protecting wild horses and burros can be traced back to animal rightists. Far fewer mute swans and feral cats are being killed due to the intervention of animal-rights activists. These organizations have millions of members and plenty of clout.

The mere presence of such powerful animal-welfare groups can have a chilling effect on the management of non-native animals. "People at the Conservancy have been a lot more nervous about making management decisions since this business with PETA," said Naehu. Holt added, "There's no doubt that some agencies and organizations that don't feel that they are well prepared to weather a protest are opting for less controversial approaches, including doing nothing."

In their OTA report, Stanley Temple and Dianne Carroll write, "Many exotic game animals can only be killed under license and during a carefully enforced season designed to ensure the species' continued existence." As the case of wild pigs in California shows, fish and game departments and their constituents can at times rival animal-welfare groups in their ability to protect alien species. Fish and game departments do seem to feel concern for these animals, perhaps partly out of parental instinct. Temple and Carroll note that fish and game departments imported, nurtured, and spread many exotics. One program alone brought millions of game animals, at least 32 species, to the United States from 1948 to 1970.

Social and political impediments notwithstanding, the largest obstacles to the control of exotics are biological. Even if anti-pig forces were free to use every possible means to annihilate *Sus scrofa,* the pesky critters wouldn't vanish anytime soon. Nor would they vanish anytime later. Eradicating pigs from a large area—say, California—would be astronomically expensive, if not physically impossible. Studies indicate that at least 70 percent of a population of these peerless reproducers would have to be killed annually just to significantly lower pig numbers. The chances of eliminating a less prolific species aren't much better, according to Temple and Carroll. "Unless an eradication program begins early and is intensive," they write, "only a reduction in numbers is likely to result." They add, "Efforts to eradicate exotics that have already become established in the wild have not generally been successful, except on a very local level."

That final phrase contains the seeds of a strategy for dealing with wild pigs a well as with some other species of exotic animals. Reg Barrett, the pig expert from the University of California, Berkeley, thinks pig management should be adapted to fit different land uses. Natural areas, for example, ought to be free or nearly free of pigs. In the late 1980s, Barrett led a project that demonstrated the feasibility of eliminating pigs from a state park. It wasn't easy, but two years of trapping, fencing, and hunting with dogs removed every single pig. However, only 135 pigs had inhabited the park, it was only about 4,500 acres in area, and the terrain was fairly open. For those with smaller budgets, larger properties, or more difficult terrain, Barrett recommends keeping the pig population at an acceptable level through a mix of control methods, especially sport hunting. Others suggest the same approach to address the economic concerns: Keep all pigs out of small, high-value areas, such as vineyards, and merely suppress pig populations on pastures and other less valuable sites.

Even such a modest approach hasn't gotten far yet, held back by Fish and Game's priorities and tight budgets, and the general lack of awareness regarding the destructive nature of invasive exotics. But one can take relative solace in Brian Hunter's words: "Pigs aren't the worst exotic in California." Sadly, more destructive alien invaders inhabit the Golden State and the rest of the nation.

CHAPTER 3

THAT CHEATIN' HEARTLAND

No one knows exactly how it got to the United States. Probably it entered in shipments of grain seeds. No one knows exactly how it spread, though it often travelled by train, getting swept out of stock cars to the waiting earth along the tracks. Few people notice it even when they walk through a field of this thin-stemmed, shin-high grass, except to curse the bristly awns that invariably hook onto their socks. But when someone picks an awn from his sock, he's holding between his fingers an invader that has changed the nature of the nation's sagebrush grasslands.

And plants can change a place forever. Plants form the foundation of a biological community; they are the center from which the food web fans out, and they profoundly influence the soil, the water, the air, the very shape a community takes. When exotic plants replace native plants, the habitat changes. If one or more of those exotics is an aggressive, dominating species to which the ecological community is susceptible, then the system will be utterly transformed, and it's not likely to be a change for the better.

Recent surveys indicate that at least 6,000 non-native plants have become established in the United States. Most are benign or of minor consequence, but hundreds cause serious trouble. These are known as exotic invasive weeds—a weed simply being any plant that humans deem to be in the wrong place. The harm done to agriculture by alien weeds has long been recognized. According to the 1993 OTA report on invasive species, they cost U.S. farmers $3.6 to $5.4 billion a year in crop losses, soak up about two billion dollars worth of herbicides annually, and comprise a large majority of the crop weeds assaulting our nation's cultivated fields. But farm weeds are not the only concern. Even more worrisome in the long run is the ability of invasive plants to transform entire biological communities.

All too many exotic plants behave as botanical barbarians, laying waste to the countryside. An exaggeration? Talk to someone in the southern United States about kudzu. This Asian import sweeps over woods like a green tidal wave, blanketing the ground, the shrub layer, even the treetops, until the entire community has been engulfed. Chat with farmers or hikers in the Midwest, where multiflora rose has turned fields and natural areas into thorny thickets. Converse with a conservationist in the Southwest, where salt cedar has usurped stream banks from native willow communities and then proceeded to suck scarce water from these desert lifelines. Or talk to someone in the intermountain West about cheatgrass.

Cheatgrass is the scourge of the sagebrush grasslands, the plant whose awns hook onto socks. Its scientific name is *Bromus tectorum,* but it got the name "cheatgrass" from farmers who felt that the pernicious weed cheated them of their crops. People also call it downy brome, downy cheat, and bronco grass. This profusion of common names indicates the invader's long geographical reach.

Cheatgrass first showed up in the sagebrush grasslands in the 1890s. These grasslands define the intermountain West, roughly four times the size of New England. The arid region stretches west from the Rockies to the Cascades and Sierras and north from central Utah, Nevada, and northeastern California to Canada. Those first sightings occurred in Utah and Washington. By the turn of the century, people had spotted small cheat infestations in much of northern Utah, eastern Washington, and eastern Oregon; by 1915, cheat had appeared in parts of southern Idaho and northern Nevada; and by 1930, it occurred throughout the intermountain West.

But this was only the first stage of cheat's conquest of the region. Though by 1930 cheat had shown up all over the intermountain West, it was sparsely distributed, usually just a few specimens scattered amid a sea of native plants. In the second stage, rather than expanding the boundaries of its territory, cheatgrass grew thicker, building its populations, turning those scattered cheat settlements into an urban mass of *Bromus tectorum*.

Sometimes at a remarkable speed. "The spread was often so rapid as to escape recording," conservationist Aldo Leopold wrote in his 1941 article "Cheat Takes Over." "One simply woke up one fine spring to find the range dominated by a new weed." Cheat is now the most common plant in the intermountain West, and cheatgrass populations continue to grow in density. Thousands and tens of thousands and hundreds of thousands of acres of our grasslands have nothing but cheat growing on them.

I first encountered cheatgrass in the summer of 1992 while driving through the Snake River plains of southwestern Idaho. Vast tracts of sagebrush steppe fanned out beyond the windshield. A few rounded hills and a horizon of mountains hazy blue in the distance underscored the prevailing flatness. The silvery green of sagebrush flashed here and there, but for mile after mile nearly the entire landscape was painted the straw yellow of dry summer grass. And virtually all of it was cheat.

In Boise I met up with Mark Hilliard, a wildlife biologist with the Bureau of Land Management (BLM), the federal agency that oversees most of the rangelands in southern Idaho. Hilliard took me out to the Snake River Birds of Prey National Conservation Area (NCA), a vast chunk of the Snake River plains a few miles south of Boise. We stopped in enormous stands of pure cheatgrass, thousand-acre seas in which wispy stalks of dry grass swayed in the wind. In monocultures such as these, researchers have counted as many as 10,000 cheat plants per square meter.

I asked Hilliard about life without cheatgrass in the Snake River plains. He said there are a variety of community types; one that was once common was defined by bluebunch wheatgrass and Wyoming big sagebrush and included bottlebrush squirreltail, six-weeks fescue, Sandberg's bluegrass, and perhaps a couple of dozen other species. Between the bunchgrasses, which grow in clumps separated by open ground, one finds a thin, gray-black coating over the soil. This is the cryptogamic crust, a living veneer of lichens, mosses, and other organisms that shuts out infectious agents.

I later asked Hilliard if we could go to a site where this rich native community was still intact. My seemingly simple request prompted some head-scratching. Not that Hilliard wouldn't be happy to take me; it's just that he couldn't readily think of any site nearby, so complete has been the transformation of the Snake River plains. Finally he thought of a place. We drove maybe six miles to what was the only semipristine native site for many miles in all directions. It turned out to be a mere two acres, a tiny island of native vegetation surrounded by cheatgrass. Hilliard said that the little haven had resisted invasion because a fence had been around it for a long, long time. A rail fence kept cheatgrass out? No, replied Hilliard, the fence kept cows out.

Cheatgrass flourishes amid ruin. Even though cheat found the environment of the sagebrush grasslands agreeable, an undamaged range could have withstood or at least greatly slowed down the invasion—most of the widely scattered cheatgrass seeds would have been kept away from the soil by the cryptogamic crust. Unlike cheat, many native plants have special awns that drill through the crust, allowing their seeds to reach the soil. Those cheat seeds that did find entry via a gopher mound or a burn would have had difficulty competing for resources with the thoroughly established plants of the native community. In the years following its introduction, cheatgrass would not have spread so successfully. But when the alien cheat first invaded the neighborhood in the late 19th century, it found that all the doors had been left wide open.

Since the most recent ice age, the plant communities of the

intermountain West had evolved without being subjected to frequent pressure from herds of large herbivores. Bison hadn't thronged this region as they had the Great Plains. But when domestic livestock (all non-native species) were introduced, native plant communities suffered because they were unaccustomed to grazing, trampling, and wallowing. Due to their heft, cattle hammered the sagebrush grasslands, and the hooves of the cattle shattered much of the cryptogamic crust, providing openings into which alien cheat seeds could easily settle.

Additionally, grazing livestock eat the native bunchgrasses, removing the leafy material needed for photosynthesis. These indigenous grasses are especially vulnerable to grazing because their meristems (growth points) are located near the tips of their shoots. When a cow munches a bunchgrass meristem, that shoot may not bud and produce seed that year. And livestock can also spread cheat seeds—in two ways: the awns catch in the animals' hair and drop out who knows where; and because cattle will eat cheatgrass, which is palatable for a brief period in the spring, they spread cheat through their feces. A study found that seeds may be retained in the digestive tracts of livestock for six to ten days. Another study discovered that, in one grazing season, a single cow on average will disperse 900,000 viable seeds from 36 weed species.

Cheatgrass, in contrast to the natives, evolved for thousands of years under pressure from camels, horses, and other formidable grazers, and can readily coexist with cattle. As long as cheat has time to set seed, which it does quickly and prolifically, it isn't affected by trampling; as an annual, the adult cheatgrass plant becomes dispensable once it has done its reproductive duty. Cheat, unlike native bunchgrass, can stand up to repeated grazing because cows usually bite above its ground-level meristem, which allows the plant to grow back.

Once cheat has established colonies within a native plant community, it reveals a decisive adaptation: Cheat can germinate and grow rapidly. This ability gives cheat a conclusive edge when its seedlings are pitted against native seedlings. This is the kind of trait that often separates major invasive exotics from minor exotics. Several close relatives of cheatgrass also showed up in the intermountain West, but they didn't have quite the right stuff to pull off the sweeping hostile takeover that cheat managed.

Typical of the unseen battle between cheatgrass and native plants is cheat's interaction with bluebunch wheatgrass, which was the most common grass in most of the intermountain West before the ascendancy of *Bromus tectorum*. Both cheatgrass and bluebunch wheatgrass seeds usually germinate in the fall. Bluebunch roots grow perhaps six inches, and then growth slows to a near standstill when winter cools the soil at the depth of the seedlings' root tips to temperatures below 48 degrees. But cheat roots can grow at much lower soil temperatures, at least as low as 37 degrees. Cheat roots also grow about 50 percent faster than bluebunch roots. By the time winter chills the top layers of the soil, where bluebunch stalls, cheatgrass roots have pushed down to soil depths where temperatures rarely drop below 37 degrees. This enables cheat to continue to grow throughout the winter.

In spring, when soil temperatures rise sufficiently for bluebunch wheatgrass's roots to resume growing, cheatgrass roots are already a foot or two long and secondary roots are often well developed. As the thirsty young bluebunch roots move deeper, they encounter almost no moisture—the cheat roots have already taken it. As a consequence, during the dry summers characteristic of the region, many young bluebunch-wheatgrass plants suffer from lack of water and some even die.

When competing with cheatgrass seedlings, the seedlings of most native plants in the sagebrush grasslands meet the same fate as bluebunch wheatgrass. But the competitive advantages enjoyed by cheat seedlings do not by themselves lead to rapid domination by cheatgrass. Native bunchgrasses are perennials, and the mature plants do not have to grow their root systems back from scratch every spring. Therefore they already have deep roots and can outcompete cheatgrass for moisture. Cheatgrass wouldn't have gotten nearly so far, so fast if livestock grazing hadn't weakened the adult bunchgrasses, which set up a contest among seedlings. Cheat can't push around a full-grown adult, but it may be able to bully a little bluebunch seedling.

Unfortunately in the latter half of the nineteenth and the first few decades of the twentieth century, livestock grazed the West hard. In the 1930s, the government conducted the country's first major range survey. "The plant cover in every range type is depleted to an alarming degree," the survey's authors concluded. "Palatable plants are being replaced by unpalatable ones. Worthless and obnoxious weeds from foreign countries are invading every type, and throughout the entire western range the vegetation has been thinned out until even conservative estimates place the forage value at less than half of what it was a century ago. . . . There is perhaps no darker chapter nor greater tragedy in the history of land occupancy and use in the United States than the story of the western range," the authors added.

Even in the parts of the West where bison had toughened native vegetation to the depredations of livestock, the aridity of the region and mismanagement of cattle and sheep still led to degraded land. In the bison-free intermountain West, land suffered worse yet. In 40 or 50 years, cattle reduced much of the region's robust grasslands to gaunt shadows of what they were formerly. Sometimes cattle left little but dirt in their wake. Or to put it another way, cattle turned much of the intermountain West into cheatgrass heaven.

The overgrazing reached such proportions that even the members of Congress from western states—Stetson-topped men whose fierce support of the cattle industry is legendary—began calling for regulation. It finally came in 1934, when Congress passed the Taylor Grazing Act. This law reined in the ranchers, but only from a dead run to a gallop. In the last decade or so, the pace of overgrazing has slowed further, perhaps to a trot. But many ranchers haven't said "Whoa" yet. Right there in southern Idaho, I saw plenty of evidence that overgrazing continues.

I visited many rangelands in the Snake River plains. Some had been chewed and trampled into eroded earth whiskered with a bit of stubble. Some sported healthy stands of grasses, but those stands consist of non-native species, such as crested wheatgrass, that are favored as cattle forage. Some sites were still in good shape, characterized by a fair number of native plant species. Little land I saw was in what I'd call excellent shape. Fortunately many rangelands are in much better condition than those in the intermountain West.

Some of the blame for overgrazed grasslands lies with the BLM and the U.S. Forest Service, the two agencies that oversee nearly all of the grazing on the nation's public lands—and most western rangelands are public. For much of their history, these agencies have been virtual subsidiaries of the livestock industry, pretty much giving ranchers free rein on the public lands they lease. However, in recent years the BLM and the Forest Service have put more effort into curbing harmful grazing practices; sometimes they've even succeeded. But often grazing reforms have run into a wall. One agency staffer once likened that wall to the former Iron Curtain: He called it the Bovine Curtain.

Through a mix of myth, inertia, and their status as an important industry in the region, ranchers still exert enormous influence in much of the West. If a proposed reform threatens the industry's ability to manage grazing as it pleases, even on public land, some ranchers are quick to phone their senators and representatives in Congress. And they usually get the results they want. A biologist familiar with grazing issues told me, "It's going to take the hammer of the courts," to force some ranchers into improving their grazing practices. A range manager went a step further, saying, "It's going to take a few funerals before we get this squared around." I'm not sure if he was referring to the passing of stubborn, old ranchers, of good-ol'-boy agency bureaucrats, or of both.

It's important to note that plenty of ranchers treat the land well, fully aware that only a healthy range will provide ranching with a healthy future. In fact, there's been an increase in environmentally sensitive grazing management in recent years. But much improvement by many more ranchers and land managers will be necessary if the sagebrush grasslands are to be even modestly protected from cheatgrass.

It seems as if enlightened management could heal the range and stop cheat from completely taking over. Bear in mind that cheat can't spread nearly as quickly if the cryptogamic crust remains unbroken by trampling herds. And again, it is overgrazing that weakens mature native plants, which could otherwise compete successfully with cheatgrass. That in turn sets up the contest among seedlings—which cheat invariably wins. And bear in mind that despite the scarcity of intact native-plant communities on the Snake River plains, many acres of the intermountain West still sport reasonably healthy native assemblages in which cheat may be present but not dominant. So stop overgrazing, let the mature native plants stand tall against cheatgrass, and the spread of this invader will halt. Right? Unfortunately it's not that simple. For there's something else about cheatgrass that makes it a terrible foe. It promotes fires like the native sagebrush grasslands never saw.

It happened in July 1995. At seven o'clock in the evening, the day was still hot, even away from the fire. Blas Telleria and his crew, the nine BLM firefighters who comprised Unit C, were extinguishing a small blaze along the freeway not far from Boise. His unusual name reflects Telleria's Basque heritage; Basque sheepherders settled in southern Idaho in the nineteenth century, and their descendants still constitute a vibrant cultural community. Like many professional firefighters, Telleria battles blazes in the summer and works elsewhere the rest of the year. (He teaches American government, economics, and psychology at a high school in Boise.) Telleria, who has been fighting fires since 1966, was the Unit C crew leader that summer day.

At about 7:30 the dispatcher called Unit C and sent them down to another fire, this one out in the Snake River Birds of Prey National Conservation Area, not far from the farm town of Kuna. Telleria and his crew drove their three pumper trucks out to what is now known as the Point fire, where they joined another BLM crew and a volunteer crew from Kuna.

"When we arrived it wasn't a big fire," said Telleria. "It's rather docile, in the neighborhood of a 150 acres. The smoke is going straight up," described Telleria, which indicates calm conditions and simple firefighting. The cheatgrass formed a dense understory beneath a canopy of sagebrush, and the fire was just creeping along through the cheat.

Over the radio the firefighters received a warning to keep an eye on a thunder cell to the south. "As it comes to pass," said Telleria, "that thunder cell does come within the boundaries of the fire. It starts to dissipate with real high, high winds, 45 miles an hour, and it blows this fire up." Telleria told me the effect is similar to opening a shower door, which makes the steam roll out at the bottom. The water from the cloud pushed the air down, creating a down draft. However, because the weather was so hot, the rain evaporated before it could reach the ground to help control the fire.

"As this blowup happens," said Telleria, "our engines are scattered all around the fire perimeter in a mop-up stage. We'd pretty much stopped the fire from spreading. Then the wind hits. It rolls embers, sagebrush, out into the unburned fuel. It ignites the brush here, it ignites the brush there, and because it's cheatgrass, very dry fuels, and high winds . . . it accelerates very, very rapidly. This fire is probably moving in the neighborhood of 20 miles an hour or greater. It's scooting right across the ground. And the reason it scoots across the ground is cheatgrass, because it is volatile. It allowed the fire to move as fast as the wind could push it.

"In a strong storm like with the Point fire, where you have such high winds, the heat isn't going straight up like it does in most fires. The heat is blowing around at ground level, preheating the fuel in front of it. All that heat that would naturally dissipate up, like a campfire, instead is taken with the wind. It takes the moisture out of the fuel. It moves the conductive heat and the radiant heat up through that unburned grass, and it just explodes."

Telleria said that soon after the storm hit, the firefighters' radios filled the airwaves with reports like "there's fire behind you" or "the west flank is going." "Pretty soon people start moving into the black," said Telleria. "They're getting into the burned areas because they sense danger. You always know in the back of your mind what your escape route is and you always hope that it's ten feet away. So most all those crews moved into the black, the safe areas. In a matter of maybe five minutes, this sleepy little fire had exploded to about five hundred acres, a huge change.

"Unfortunately for some reason one of the Kuna engines found itself out in front of the fire, with a stalled engine. The way that wind is laying that radiant heat along at ground level, preheating the fuels, blowing more heat way out in front of it, smoke and ash in front of that, the fire is going to overrun them far faster than . . . ." Telleria paused at the memory, then continued, "It was fast. It's going a minimum of 20 miles an hour, and what complicates it is that the smoke and the ash are blowing way out in front of the actual flames at 45 miles an hour.

"I was driving down a road trying to get out in front of the fire, to some sort of vantage point, when I saw their engine sitting [about 700 feet from the road] out in this pile of brush and cheatgrass with this fire coming at them. I desperately tried to make contact with them, but in retrospect we learned they were not on our channel. As I travelled down the road, a finger of fire, right off the road, was moving along with me. Imagine you're looking through the frame of a window. In the left part of the window you see the flame coming very rapidly. In the middle of that window you see this engine out there and you know that this flame is going to go across that window frame real quick." Telleria said he couldn't tell if any of the Kuna firefighters were in the engine; he wondered if they might have abandoned the stalled vehicle and run for it.

"It was just seconds before the flames surrounded them," said Telleria in a quiet, even voice. "There was another finger of fire just on the other side of the engine. It was god-awful ugly. It was like the devil's hand reaching out there. It just washed right over the engine." Moments later the hellish vision was blotted out. "The smoke and the ash start pelting us," said Telleria. "I can't see, I can't open my eyes, it gets real, real warm. The fire storm rolls over our engine and our crews. Fortunately we were on the edge of a road. In my vehicle it turns dark like an eclipse, and then all of a sudden it turns bright orange." Telleria and the people with him survived because they were on the road and on the outskirts of the fire. "We hunkered down," said Telleria. "We knew something bad has happened. In a flash the fire passes us, but we still can't see because of the smoke and ash swirling in the wind."

When the miasma thinned, firefighters rushed out through the glowing stubs of the sagebrush to the stalled engine. Their fears were realized when they found the burned bodies of two of the Kuna volunteers. "It weighs heavy on my soul all the time," said Telleria.

Cheat fires rarely result in human deaths, but they are playing a major role in the gradual demise of the sagebrush grasslands. Fires are a natural force in the grasslands, but before cheat they were infrequent and small. Because bunchgrasses grow in scattered clumps and the cryptogamic crust doesn't carry fire, fire doesn't spread readily in a native bunchgrass community. Many native plants stay green well into the summer, which delays the onset of the fire season. No one is sure exactly how often fire hit native communities in pre-cheat days, but a widely cited 1990 study by Steve Whisenant, an ecologist then at Brigham Young University, indicates that historically any given site in the Snake River plains burned about once every 60 to 110 years.

Cheatgrass, though, because it provides continuous fuel, carries fire with frightening ease. Cheat usually dries out by mid-June, some two to four weeks earlier than most of the natives, significantly lengthening the fire season. In most cases, its fine growth readily ignites. Once cheatgrass fills in enough of the open spaces in a bunchgrass community, the site is virtually certain to burn soon—on average, every three to five years according to Whisenant's study. Cheat increases the size of fires, too. In a pristine Snake River plains–bunchgrass system, if one existed, an average fire might burn 50 to 100 acres. In an average year in the Boise district two or three decades ago, with cheat already established, 60 to 70 fires burned a total of 25,000 to 30,000 acres. In an average year in the mid-1980s, after several more turnings of the cheat-wildfire cycle, 115 fires burned 186,000 acres, despite increased suppression efforts. Bill Casey, district fire management officer for the lower Snake district of the BLM, notes that typically more acres burn in his district than in any other district in the lower 48. Incidentally, the Kuna Fire eventually burned more than 10,000 acres.

The speed of cheat fires makes them difficult to contain; the Kuna Fire was not rare in this regard. Everyone in cheat country seems to have a story about an extreme example—like the one I heard from Stephen Monsen, a botanist with the U.S. Forest Service. One day Monsen was driving through southern Idaho when he caught up to a cheat fire that was blowing parallel to the highway. Monsen slowed down to pace the fire, but he didn't have to slow down much—it was sizzling along at about 40 miles an hour.

What Casey calls "multiple occurrences" also make it difficult to fight cheat fires. Dry thunderstorms rumble along the storm track from Oregon and unleash thousands of lightning bolts into cheat stands throughout the Snake River plains. Often many widely separated fires start within a few hours; at one point in 1986, 46 fires were burning simultaneously in the Boise district. "It's a zoo," said Casey. "We're absolutely inundated with fire reports."

Humans start many fires on the plains, too. Maybe someone's off-road vehicle throws a spark or maybe a careless smoker flicks a lighted cigarette out the window of his car as he streaks down Interstate 84. In 1996, an off-duty police officer allegedly started a 12,000-acre blaze in the cheat-ridden foothills above Boise by engaging in some target practice using tracer bullets. As of 1998, the case was still in court.

After a burn, cheatgrass usually flourishes, even if it previously formed only a part of the plant community. In a severe blaze, such as the Point fire, the sagebrush and many other shrubs are killed and the aboveground growth of the herbaceous species, including cheatgrass, is consumed. After the crowns of mature native bunchgrasses burn, their roots wither; and with the mature natives incapacitated, the competitive advantages that cheat seedlings have over native bunchgrass seedlings enable them to win the recolonization race.

Less severe fires may spare mature native plants, but if the fire stresses them too much, and especially if subsequent grazing adds to their stress, they become less competitive with cheat. That allows cheat to proliferate, which increases the odds that the site will burn again in the near future. That next burn will either kill the remaining mature native plants or stress them further, aiding the spread of cheat and setting the stage for the next fire. This downward spiral generally continues until cheat dominates the community. Fire begets cheat and cheat begets fire.

Generations of ranchers have suffered greatly from their predecessors' overgrazing practices, which established the destructive cheat-fire cycle. Most of the year, cheatgrass provides cattle and other livestock with poor forage—dry, barbed, almost unpalatable, and rangeland hazards—sharp awns stab stock, causing lumpy jaw (abscesses caused by trauma and subsequent infection) and blindness. Back in 1992, I drove out into the Snake River plains to visit with Bob Baker. As a rancher whose range has its share of cheatgrass, he tries to make the best of a bad situation by taking advantage of cheat's early springtime flush, when it is abundant and palatable. Contrary to conventional wisdom, he even got some grazing value from cheat during the winter, but as the fires escalate and cheat swallows more and more of the landscape, Baker and other ranchers worry that they could be left with little but cheatgrass, which creates additional problems.

Ranchers like variety for good, reliable year-round forage—"like having meat, potatoes, and vegetables on your plate," said Baker. Also, with little besides cheat, ranchers have a devilish time trying to decide stocking rates. "I'm sitting here with X number of cows," said Baker hypothetically, "and I'm relying on that winter range to be there when they get there the first of November. But we could have a big fire and the forage could be gone, and then what do I do?" Baker bought his own firefighting equipment in order to pounce on blazes that start on his land, which happens about once a year. "You can lose all your fall and winter forage in an afternoon," he says.

Following up with Baker five years later, he said he was getting along all right, but year by year cheatgrass was making his life harder. His firefighting equipment had turned out to be pretty much useless.

He said the BLM had beefed up their firefighting efforts, but they usually couldn't get out to his spread in time to do much. "The fire just goes so fast, before they can respond. And if there's a wind with it, which normally there is with a thunderstorm, it can take out a lot of acres in a hurry." He reports that maybe 200 to 1,000 acres of his 17,000 acres of winter range burned each year—and he considers himself lucky. "The reason we haven't had a big fire is that most of our winter range is intersected by county roads and cut into smaller pastures. Even though we lose one pasture, we don't lose the whole thing, so we're fortunate in that way. Last year our neighbors to the south, their land's all in one big block, and they lost 20 or 30 thousand acres, which was a sizeable percentage of their winter range. They didn't have the roads for firebreaks."

Even with his dirt-road firebreaks, Baker figures that cheat already has overrun more than half his winter range. "An awful lot of our winter range has now become predominantly cheatgrass. So we supplement more with protein supplements and such, but the livestock don't do as well. It's just becoming a monoculture of cheatgrass out there." And then there's his summer range: about 12,000 acres up in the hills. "In '94, the big foothills fire burned our entire summer range," he reports, "and a lot of other people's, too. Burned more than a quarter of a million acres." Much of the burned area, including his summer range, had boasted a fairly healthy component of native plants, but it was laced with cheat. The fire had started lower down in pure cheat lands, "but the fire just came on up to us and kept on going, went on through us. It took trees, it took the perennial grasses, it took everything." Baker, the U.S. Forest Service, and the BLM were able to rehabilitate his range—largely because it occupied a relatively wet site—but it cost Baker and the taxpayers plenty in time and money. Baker adds, "We had to go out and rent other ground for two years while that rehabilitated." If ranchers and agencies could summon the will and the wherewithal to follow the lead of range managers who rehabilitate deteriorating lands before they burn and succumb to invasive weeds, their actions would save a great deal of native habitat—and, in the long run, a lot of money.

Cheatgrass diminishes forage for wild animals, too. Elk and deer descend from the snow-bound mountains north of the Snake River plains to overwinter in the foothills, where they escape the worst of the cold and browse on sagebrush and bitterbrush. The availability of these native shrubs in this narrow wintering belt is crucial to the survival of these animals. But cheat fires are burning more and more of the foothills and burning to higher and higher elevations. As an annual, cheat adapts quickly, and adapted strains are stealing into habitats that cheat originally found inhospitable: It has taken over the understory of much of the piñon pine-juniper zone and is even turning up in ponderosa-pine forests at elevations as high as 7,000 feet. The cheat-fire cycle has degraded vast swaths of winter range; during harsh winters, desperate deer and elk are forced to seek forage on lower elevations, which no longer are mantled with the shrub communities that for centuries carried these animals through such winters. At these times the animals often plunder farmers' fields, defoliate backyard shrubs, and collide with vehicles.

The impact of the cheat-fire cycle ripples out to affect wildlife in ways less obvious than the debasement of elk and deer winter range. A particularly distressing example is occurring in the Snake River Birds of Prey National Conservation Area (NCA). On a warm May morning in 1992, some companions and I headed up the Snake River in a boat from Swan Falls Dam. Above me the strong high-desert sun ignited the bold reds, browns, and golds of the sedimentary bands that stripe the 700-foot-high basalt cliffs of the Snake River Canyon. After a while, we spied a bird as it suddenly bulleted over the canyon rim: a male prairie falcon. These streamlined birds of prey can scream along at 150 miles per hour, and this fellow had his throttle up pretty high. He whizzed past a female prairie falcon perched on a ledge, drawing her into the air with him, where they began a courtship ritual. We cut the boat motor and watched quietly. The male climbed high above his cruising mate, then jackknifed and whistled downward. Wings tucked, he streaked by her, a brown-and-white blur. He flashed upward and rocketed past her several more times. It was like watching the Blue Angels thunder overhead at an air show, except that the falcons maneuvered silently and with far more grace and agility. After his series of power dives, the male pulled alongside the female and they curveted in tandem, banking, rolling, arcing upside down in great parabolas. As I gazed skyward, I understood viscerally the reason that people had clamored for protection of this area.

The NCA harbors the greatest concentration of nesting raptors in North America. Some 5 percent of the world's prairie falcons hatch and fledge their young there. Golden eagles, ferruginous hawks, long-eared owls, and ten other raptor species likewise find the area a fine place to raise a family. On average, about 700 pairs of raptors nest in this sanctuary—a huge number given the solitary, roving nature of birds of prey. In addition to the nesters, peregrine falcons, northern goshawks, bald eagles, and several other raptor species migrate through the NCA or take up temporary residence there for part of the year.

The Secretary of the Interior established the NCA in 1971, designating 26,000 acres of the river canyon, including the stretch where I watched the courting prairie falcons, off limits to many forms of development or intrusion. Happily, the nesting area was protected, but the feds had overlooked something essential. The ideal nesting ledges and crevices on the canyon cliffs alone do not make this area raptor heaven. The sagebrush grassland that embraces the canyon completes the equation. Out there amid the shrubs and grasses live the prey that the raptors must have for themselves and for their ever-hungry young. So in 1980, the Secretary of the Interior expanded the NCA to include nearly half a million acres of that prey habitat.

The bulk of the 1980 addition lies northeast of the canyon rim. The expanse is blessed with a mantle of deep, fine soil beloved by several species of burrowing rodents, notably Townsend's ground squirrels. It so happens that these rodents proliferate during the time when raptor parents need tons of food for their demanding offspring. Prairie falcons in particular depend on Townsend's ground squirrels.

Black-tailed jack rabbits also thrive in the native habitat of the 1980 addition. These hares are too big for some birds of prey to manage, but jack rabbits are the mainstay for the sizeable population of golden eagles. With a wingspan of about seven feet and hand-size talons that grip with a force of 2,000 pounds per square inch, golden eagles can handle a jack rabbit as easily as a person can pull a can of soup off a supermarket shelf.

But the goods on the shelves at the raptors' market are beginning to get sparse. The one-two punch of the cheat-fire cycle has knocked out most of the native grasses and seeds that Townsend's ground squirrels favor, and cheatgrass doesn't work as a replacement. As cheat has spread across the NCA, about two-thirds of which is now dominated by *Bromus tectorum,* squirrel numbers have been dropping. Not surprisingly, so have prairie falcon numbers. Probably other birds of prey also have suffered from the decline of the ground squirrels. Black-tailed jack rabbit numbers have been dropping as well. Jack rabbits need a place in which to conceal their rangy bodies and six-inch ears from the sharp eyes of predators, and a stand of cheatgrass just won't do. The black-tails prefer to hide in sagebrush, but the cheat fires have burned off most of the sagebrush and repeated burns keep it from coming back. The golden eagle population is dwindling and female eagles seem to be laying fewer eggs. Again the ecological ripples have fanned out through the food web.

Steve Jirik, a rangeland management specialist for the BLM in the NCA, works to restore prey habitat. He points out that the primary mission for the Snake River Birds of Prey National Conservation Area is to conserve birds of prey—and that mission isn't going well, thanks to the cheat-fire cycle. The BLM and others have been trying to deal with cheatgrass in the NCA, but they have had few successes. Ten years ago, the agency tried greenstripping: They planted firebreaks with vegetation that would resist fire and invasion by cheatgrass. Initially greenstipping flopped, due to an extended drought and the decision to allow grazing before the resistant vegetation had become established. In fact, the project just made matters worse; cheat and other weeds overran the greenstrips, creating corridors of highly flammable material. In recent years, fortunately, the BLM has been able to reclaim most of the greenstrips, which should help slow the spread of cheat fires in the coming years.

The agency also had to persevere through early failures in its efforts to rehabilitate cheatgrass burns. "We tried disking, plowing, and other mechanical methods to reduce cheatgrass competition prior to seeding," said Jirik. "But these methods obliterate any remaining native plants, especially perennial grasses. We obliterated the cryptogamic crust, which still was generally okay where sagebrush had been. We finally realized that if you're not going to be able to control the [regrowth of] cheat, then it's not worth wasting the taxpayers' money by trying to rehabilitate a burn."

In the last year or two, researchers, land managers, and ranchers battling cheat have found renewed hope. This in part is due to BLM's new methods: reestablishing native plants on sites that have burned but have not yet been engulfed by cheatgrass. The BLM has had some luck restoring sagebrush if they seed it before cheat begins its assault in the spring following a fire. Seeding sagebrush is quite a turnabout for the agency. For years, at the urging of the ranching industry, the BLM *removed* sagebrush in order to expand the grazing capacity of land.

In addition, the BLM has found some exotic grasses, mainly Siberian wheatgrass and Russian wild rye, that grow well in the driest sections of the NCA. "There's a lot of demand for us to put in native grasses," says Jirik, "but the problem is that in much of the NCA, we're down to the seven-to-ten-inch precip zone, which is real marginal for successful seedings, and at this point there just aren't many native perennial grasses available to us which will compete with the cheat and persist in this area." Researchers are working on developing native-seed sources and methods of reestablishing native communities, but in the meantime, as Mark Hilliard once said, "You've got to stop the hemorrhaging before you can start the curing." In 1996, 50,000 acres burned in the NCA despite aggressive fire suppression efforts. That's about 14 percent of the total conservation area. "The main thing," says Jirik, "is to try to reestablish a perennial grass-shrub community and break that fire cycle."

Planting exotics to combat an exotic may raise some eyebrows, but the other major source of hope in the cheat war raises hackles. It's name is Oust. It's a new herbicide, a chemical, and that upsets the many people who worry, with ample reason, about our nation's addiction to chemical remedies for weed problems. (Weed control, mostly on farms, accounts for well over half of all pesticide use in the U.S.) But desperate land managers in cheat country are excited. Early tests indicate that, after a burn on a cheat-infested site, an application of Oust will keep the cheatgrass from growing back for a year or even longer. Research further indicates that Oust has only a minimal effect on any native perennial grasses that may still inhabit the site. However, Mary O'Brien, a botanist and internationally known critic of pesticide use, says of Oust, "It's just terrible [due to its potency]. For a whole acre, you use the equivalent of a little packet of sugar in a restaurant. It can blow for miles, land on a fruit tree, cut back its reproduction, and you can't even detect that Oust is there." Thomas Pfleeger, an Environmental Protection Agency plant physiologist who spent several years studying the family of chemicals that includes Oust, confirms that it has such drawbacks.

Whatever the side effects of Oust, it appears that spraying Oust on the mat of cheatgrass duff and seeds that remain after a fire, combined with the seeding of desirable plants, will allow the perennial grasses and shrubs to become established, at which point they can outcompete cheat. But Oust is not cheap. Jirik says it costs the BLM about $25 an acre for the herbicide and its application. Due to the expense, the agency was able to treat only 11,000 of the 50,000 heavily infested acres in the NCA that burned in 1996. But if all goes well, that's 11,000 acres of restored grassland.

Dangerous fires, a drop in plant biodiversity, reduced cattle grazing, harm to wildlife—these are some of the most apparent problems caused by the cheatgrass invasion. Other problems that aren't so apparent undoubtedly exist. Fires alter the amount and kinds of nutrients available to sagebrush-grassland organisms. In particular, it's likely that repeated blazes create a shortage of nitrogen, one of life's key elements. Recent research suggests that cheat fires may damage or kill mycorrhizae, a mutually beneficial network of fungi and plant roots found in the soil. Many plants, including sagebrush, fare poorly in the absence of mycorrhizae. And who knows what else? We know relatively little about the ecology of this ecosystem, about the ways in which a major change in vegetation and fire regime may affect it. But we know enough to worry that the cheatgrass-fire cycle is leading to a fundamental, perhaps irreversible, degradation of the sagebrush grasslands.

That's the bad news. The worse news is that exotic weeds are just as nasty as cheatgrass are on the march throughout the nation's grasslands and forests.

CHAPTER 4

STAR OF GARDEN CREEK RANCH PRESERVE

S tern cliffs of dark basalt hovered over me. To the sides, the landscape sliced up and down like the patterns on a disturbing EKG, sharp ridges and deeply incised ravines alternating for as many miles as I could see, and it was a good many miles. Below, the terrain hurried downward to the depths of a steep canyon, its bottom out of sight. Dramatic scenery, but of no interest at the moment.

I continued wading about the hillside, accompanied by Janice Hill, the stewardship ecologist employed by the Nature Conservancy to look after this land, and Lynn Danly, a BLM rangeland manager who helps coordinate activities at the Tri-state Demonstration Weed Management Area—a sprawling patchwork of public and private properties that includes the land we traversed, the Garden Creek Ranch Preserve. Suddenly Hill called out. We hustled over to where she stood. Excitedly she pointed out one of the rarest plant species on the preserve—Spalding's Silene. Most rare plants growing here are varieties, but Spalding's Silene, named after an early missionary, is a full species. Hill urged me to feel the leaves. To my surprise they were sticky.

This rare plant and the other plants around us on the preserve were all native species. Hill described the habitat as a typical Idaho fescue community. Like bluebunch wheatgrass found in the NCA, Idaho fescue is a perennial bunchgrass. Sprays of unruly bluebunch stuck up in considerable numbers, and the spaces between the bunchgrasses teemed with less dominant, indigenous plants. Hill and Danly bowed down and ran their fingers over the bounty of the grassland, speaking the names of the plants as one might speak the names of friends and relatives while scanning the pages of a photo album: Arrowleaf balsamroot; Deerhorn; Hairy Albert, a lupine. Hill knelt beside some yellow bell and delicately turned up its hanging-bell flowers. Danly pointed out the pink blossoms of Snake River phlox. Hill noted some stoneseed and said that local Indians had used it for birth control. Hill and Danly figured that two or three dozen native species plus a fair number of lichens and mosses inhabited this rich grassland slope, perhaps 50 to 60 species altogether.

The Nature Conservancy began buying property for this preserve in 1988. Today the preserve encompasses about 14,000 acres of relatively healthy land, managed jointly by the Conservancy and the BLM. The preserve lies on the Idaho side of the Snake River canyon, about 200 miles north as the falcon flies of the Birds of Prey National Conservation Area. Garden Creek Ranch Preserve is a somewhat cooler, wetter place than the Snake River plains that's embedded in the Tri-state Demonstration Weed Management Area, which also includes land in Oregon and Washington.

Hill, Danly, and I had entered the preserve from the top, Danly in one vehicle and Hill and I in another. Following the lead of Corral Creek, we eased in low gear down the trail along the north bank of the watercourse. Our descent began at nearly 5,000 feet in a forest of Douglas fir sprinkled with grand fir and ponderosa pine, but even there an understory of bunchgrasses foreshadowed the predominantly grassland character of the preserve. As we headed down the steep road, we soon saw penstemons, buckwheats, paintbrush, and other grassland species on the hillside. Hill noted that the native grassland communities are considerably more diverse than those of the forest, and the riparian community framing Corral Creek exhibited considerable diversity as well. Mountain alders, dogwood, water birch, elderberry, black hawthorn, and numerous other species of trees and shrubs created a pleasing tangle around and above the narrow stream. We especially admired the 15-foot-tall syringa—also known as mock orange, because its abundant white blossoms resemble those of an orange tree. Hill reached out her window and snagged a short syringa branch laden with the small, soft blooms. I breathed in the sweet fragrance and immediately understood why Idahoans had made this plant their state flower.

Seduced by the syringa in my hand, I commented on how handsome and vital the riparian area looked. It's a common mistake to equate green with health, and Hill politely paused for a moment before disabusing me of this notion. She allowed that the overstory was in decent shape, but the native understory plants had been replaced almost entirely by exotics, including orchard grass, burdock, hound's-tongue, and our old friend cheatgrass.

Shortly after my gaffe, I spotted many more exotics on the hillside above the road. First came the ten-foot-tall skeletons of Scotch thistle, killed by the BLM's roadside spraying program. Hill pointed out some whitetop, an aggressive species that readily displaces natives. She mentioned several other species that greatly concern her, among them Russian knapweed, common crupina, and Canada thistle. These plants are among the noxious weeds that many scientists and land managers consider to be as bad as cheatgrass. Many are thorny, some are poisonous to mammals, most all are unpalatable to both wildlife and livestock; some burn even worse than cheatgrass, some have deep root systems that defy herbicides and give them a competitive advantage over many native plants: all are definitely flora non grata. Gazing out at the landscape of fine native communities scarred by pockets of noxious weeds, Hill laughed ruefully and said, "This would make an excellent study area for a class studying weeds. We've got the best of the best and the worst of the worst."

A little farther down the road we encountered the weed that

currently poses the worst threat to Garden Creek Ranch Preserve. Hill, Danly, and I clambered out of the trucks and stood staring up at a slope covered with acres and acres of yellow starthistle, a member of the aster family from Eurasia. Still greened-up after an unusually long, wet winter, the yellow star wasn't its trademark color yet, but Hill told me that in a month, when starthistle bloomed, entire hillsides would be solid yellow. The plants would grow waist-high, chest-high, even over people's heads—and they would bristle with spines that can jab right through jeans.

We stepped into a sea of yellow star. Hill and Danly spotted a little cheatgrass and a little tumble mustard (another exotic), but 99 percent of the plants we saw were yellow starthistle. At the time of my visit in early June 1997, yellow starthistle dominated about 2,000 of the preserve's 14,000 acres. Starthistle spreads at a rate of about 17 percent a year. Growing exponentially like money drawing compound interest, an infestation of yellow star can double in less than five years. I asked Hill and Danly if any native species could grow in this weedy wasteland. Good question. They set off in different directions, slowly walking, eyes to the ground. After a minute Hill called out: "I'm not seeing any native species yet, but I'm still looking." Maybe 30 seconds later she pointed at a stand of bushes about a hundred feet down the road and said, "There's some poison ivy down there. That's native."

What was perhaps the first official identification of *Centaurea solstitialis* in Idaho occurred in 1938 near Lewiston, about 25 miles north of Garden Creek Ranch. Now this weed heavily infests more than 300,000 acres in the state and in certain places it's expanding rapidly. A pair of photos taken near the lower Salmon River in Idaho suggest yellow star's virulence: the first shows a grassland free of starthistle on a bench above the river; in the second, acres of pure yellow starthistle cover the same area. BLM range conservationists took the first photo in 1983 and the second in 1993.

Sadly, the preserve doesn't have a monopoly on yellow starthistle— weeds seldom observe boundary lines drawn by humans. In Umatilla

County, in northeastern Oregon, surveys revealed that starthistle acreage ballooned from 173,000 to 286,000 between 1985 and 1992. At the other end of the state, in southwestern Oregon's Jackson County, researchers report that starthistle spread from background levels in the 1970s to 110,000 acres by 1993. Now into the steep upward curve of exponential growth, the yellow starthistle in Jackson County exploded to more than 200,000 acres in 1997. But that's nothing compared to the plague of yellow star that is rampaging across the border in northern California. In 1965, yellow starthistle already had invaded a million acres in the Sacramento and San Joaquin valleys. In 1985, surveys estimated the acreage at nearly eight million. The figure continues to climb as yellow star moves into coastal grasslands and higher elevations to the east of the valleys. This marauding invader now occupies about 40 percent of California, if you count both heavily infested sites and land that has only a few yellow-starthistle plants. But only a few for how long?

After driving a little farther down the road, Hill, Danly, and I came across Garden Creek Preserve's two-legged mountain goats, a.k.a. the survey crew. Karen Gray, a Garden Creek botanist, and two botanical technicians, Sandra Robins and Ben Flemer, were spending the summer hoofing it up and down the rugged hillsides of the preserve, mapping the vegetation, just as they had the previous summer. Researchers were working on ways to use satellite imagery to create detailed vegetation maps, but the technology wasn't ready yet. Helicopter flyovers helped, but they were too expensive to do often. So for the moment, human eyes borne by human feet were needed to find out what was growing where.

I noticed that the crew carried bear spray—aerosol cans of highly concentrated pepper extract that can repel bears by stinging their eyes, nose, and throat. Gray, Robins, and Flemer said that they often saw signs of bears, that as a matter of fact they'd encountered a large adult bear just that morning. Gray reported that she'd also had a close encounter with a rattlesnake that morning. She laughed as she described the snake dance she did, frantically high-stepping away from the viper's lethal fangs.

Such flashy hazards notwithstanding, the terrain is the crew's worst enemy. The hardy trio described the difficulties of traversing 50-degree slopes, especially given the weight and high center of gravity created by their bulging backpacks. Janice Hill shook her head and said, "I just have to laugh at the idea of those three little bodies trying to cover all that country." The significance of the rough terrain goes beyond concerns about nasty falls by the surveyors; it makes it hard to deal with the weed infestations that the crew locates. Sometimes helicopters spray herbicides on the hot spots. Helicopters cost hundreds of dollars an hour, but they can hit a lot of weeds in an hour. Managers in the Tri-state Demonstration area would like to use helicopters more often, but their current budgets won't allow it. Usually the only recourse is for some poor soul to labor over hill and dale to the site of the infestation. "You take a backpack sprayer to it," said Hill, "but carrying 50 pounds of spray and walking on these rubbly hillsides . . . well, I've done that, but I'm not doing it again." Hill says that it usually takes half a day to reach outlying weed sites. Furthermore, most of the exotic species produce seeds that remain viable for five or even ten years, so the site must be sprayed over and over. Hill would have to hire entire platoons of strong bodies to hit all the known weed sites—and her meager budget doesn't even allow her to hire enough people to do the surveying. Needless to say, only a few small, high-priority infestations get treated; most of the yellow starthistle, cheatgrass, common crupina, Canada thistle, and friends are growing without restraint.

"Spraying all the time is not the answer," said Hill. "You can use it as a temporary control, but then you need to get some vegetation in there that can compete with the yellow star over the long run." But the thought of using non-native grasses to compete with the noxious weeds sticks in Hill's craw; it runs counter to her personal passion for native plants and the Conservancy's mandate to preserve them. Still, she says, "Getting those perennial exotic grasses in there is a better alternative than having yellow star in there. But we are trying natives in some areas." Down by the river, the crew has been gathering bluebunch-wheatgrass seed, taking it to a commercial seed grower to propagate, then planting the seeds he produces on some sites they're restoring.

Hill described another control method that could replace chemical use on a few sites in the preserve. She pointed at a hill and said there was a quarter acre patch of leafy spurge up there that the preserve manager, Ralph Crowley, covered with black plastic, just as gardeners sometimes do with flower beds. After three years, it appears that the plastic works pretty well, but, said Hill with a smile, "We have had a little problem with the elk. They seem to like to slide on it. It's on a slope." The elk repeatedly climb the hill and slide back down the plastic; clearly they're playing, regardless of what some cold-blooded scientists may say about anthropomorphizing.

Wishing the crew luck and a snake-free afternoon, Hill, Danly, and I continued driving down the hill. After a few minutes, we turned northwest and headed up a narrow, steep track to the beautiful Idaho fescue community I described earlier, the one in which Hill found the Spalding's Silene. After savoring this haven of ecological integrity, Danly and I hiked over to some experimental plots where the BLM was testing ways to suppress noxious weeds with herbicides while allowing the seedlings of native bunchgrasses and acceptable exotic grasses to grow.

As we studied the plots, which were too recent to yield definitive results, I asked Danly about her long-term hopes for the hundreds of thousands of acres in the demonstration weed management area. "We're never going to get rid of the exotics," she said, "even if we do reach the pie-in-the-sky goal of getting competitive species established. The ultimate goal is to have the exotics be a very small part instead of having monocultures of them." She paused, then added, "Even stopping infestations where they are now would be a success in my mind."

Danly headed home right after our talk, but Hill and I continued on. As rain began falling, we started to worry that the steep, red-dirt side road would get too slick to safely drive on. We hurried back to the truck and Hill gingerly tested the first, moderate grade. The tires slalomed several times before we reached the next level stretch. We paused. The rain continued its tap dance on the roof of the cab. Having done the slip-and-slide routine all too often on these four-wheeler tracks, Hill decided not to risk going down the steep grades that lay ahead of us. We waited for the rain to abate, but after about 20 minutes of intermittent downpour, we realized that we were stranded. Time for the five-mile hike to the old ranch house on the river that served as preserve headquarters.

The first half mile of our trek led cross-country down the 50-degree pitch of a hill. As I sidestepped down the grassy slope, my ankles constantly on the verge of spraining, I better appreciated the hardships the crew faced. I discovered a new hazard, too. When I lightly touched a sink-sized rock for support, it peeled loose from the soft ground and began cartwheeling toward Hill. Observing standard operating procedure for hiking, I shouted "rock"—though "boulder" would have been more accurate. Hill made sure she stayed out of its path. After that we descended side by side.

Except for my attempt to crush the preserve's stewardship ecologist, the walk turned out fine. We reached the main road without further incident and headed down toward the river. Though we'd hardly been speeding in the truck, walking provided even greater opportunity to relish the richness of these grasslands—at least those that hadn't yet been impoverished by exotic weeds. We saw some mule deer and heard the wild cries of a red-tailed hawk, doubling our wildlife list, which already included black bear and rattlesnake. We also knew that elk, skunks, mountain lions, bats, white-tailed deer, lynx on occasion, birds of many sorts, and a host of other critters lived on the preserve or made use of its bounty. Providing sanctuary to these animals and the diverse plant communities that support them is the preserve's mission.

Biological diversity is characterized by three levels, all vital to ecological health. There's the species level, with which we are most familiar—moose, whooping crane, Douglas fir, coyote, western diamondback rattlesnake; the genetic level, where an individual organism may carry important genetic material that is different from that of other individuals of the same species; and the ecosystem level. Communities fall into the ecosystem category of biodiversity. It is important to have diverse habitats, such as cloud forests, cold deserts, peat bogs, coral reefs, and oak savannas. To say that Garden Creek Ranch Preserve is grassland and forest is like saying San Francisco is a city—it leaves out the details of diversity. What about Chinatown, North Beach, Nob Hill, and all the other communities that give the city by the bay its splendid variety? By the same token, the preserve consists of many distinct yet connected natural neighborhoods, such as the Idaho fescue community. Variations in soil, the steepness of the slope, aspect, microclimates, and other factors turn the landscape into a mosaic.

Hill and I walked past a gray-black, almost charred-looking community that goes by the name of buckwheat–Oregon bladderpod. Though it appeared barren, this extremely rare type of biological neighborhood teemed with life, including a suite of plants peculiar to these inhospitable habitats. We also passed a smooth sumac/bluebunch-wheatgrass community, a rarity in Idaho and probably rare globally. That morning, we'd seen another sumac-bluebunch site, but yellow starthistle had nearly obliterated it. Sometimes there are less drastic alien invasions, but invasive plants will still significantly disrupt the community: a native pollinating insect, for example, depending on one particular native plant species, will suffer when that native gets replaced by some uninvited alien; or perhaps a plant virus, which natives would keep at bay, proliferates and spreads because an invader lacks the chemical defenses to fend it off. Even expert ecologists seldom know just what ripple effects an exotic may inflict upon the intricate workings of a community.

After we'd gone a couple of miles, Ralph Crowley, the preserve's manager and resident character, picked us up in his hard-used Jeep. Cowboy hat on his head, cigarette dangling from his mouth, revolver and rifle within reach, Crowley looked like the former ranch hand he was. A longtime inhabitant of the area, he told me the history of the Garden Creek Ranch Preserve—that it had been grazed by cattle from the turn of the century until 1992, and that the cows had grazed it pretty hard in the draws and on the benches.

About supper time, Crowley reined in the Jeep in front of the ranch house, a classic of 1920's vintage: big and boxy, high ceilings, a screened front porch, and creaky wooden stairs. The decor, however, had changed now that the inhabitants nurtured plants instead of cows. Microscopes, seed bags, plant samples, topo maps, binoculars, plant presses, and other gear littered the living room and dining room. Some things probably hadn't changed, though, such as the fire burning in the potbellied stove or the wild roses in a jar that graced the kitchen table.

As we began eating instant rice, burritos, and other fare typical of isolated field stations, the conversation comprised an exchange of information between Hill and the survey crew: What plants did you see where? How many? Did that unidentified plant have expanded peduncles? Was it glandular? About the time I was on my second burrito, however, I asked them how the weed invasion affects their jobs on the preserve, and they all laughed. Hill said it's almost all they think about. "It's our main, big, top problem," she said, making sure I got the point. "It's been my job since I've been here. She added, laughing, "It's keeping me in an occupation."

The conversation then shifted to the future of the preserve. Robins, one of the botanical techs, admitted that she didn't feel too optimistic. "These [non-native] annuals have such a competitive edge against the native bunchgrasses," she said. "Right now we don't have anything that controls them. We don't have a whole lot of options." I asked her what she thought would happen on the preserve and throughout the demonstration area. "I think yellow star will take over all of the habitats it likes well, like south-facing slopes." She thinks the wetter, north-facing slopes with healthy native communities may be able to resist starthistle. "They're pretty closed systems. Every little square inch has some little plant growing in it." Karen Gray, the preserve botanist, added that the influx of cheatgrass worried her. It's filling in the spaces on many sites, providing the fuel for fires that could harm natives and lead to the expansion of cheat, yellow star, and other exotics.

Hill said that much of the work on Garden Creek Ranch focused on protecting the healthy communities: "We really want to keep those areas that are good still. That's why it's our top priority to deal with the small, satellite populations"—the beachheads of weeds that have invaded native sites. On a larger scale, the Tri-state Demonstration Weed Management Area pursues much the same goal. From the front porch of the ranch house, we could see across the muddy waters of the Snake to the burly mountains of eastern Washington and Oregon, a vast landscape still relatively free of exotic weeds. There's an awful lot of land out there that could be saved.

But saving these grasslands requires active management. Hill noted that natural succession to something other than a weedy community seldom happens anymore; the exotics maintain a stranglehold by means of fire and their competitiveness. Overgrazing threw off the trajectory, and now the proper management of the cattle alone, even their removal, doesn't cure what ails many of our grasslands. Hill lamented that just when long overdue improvements in livestock management brightened the outlook for the grasslands, trampled and trashed for so many years, exotic weeds came along and took away much of the hope.

CHAPTER 5

SPURGE OF THE GREAT PLAINS

n my rearview mirror, I could see for a couple of miles behind me. A rooster tail of dust was settling back onto the ruler-straight dirt road that led to the Fugere ranch. Home to Cindie, Ian, and Kevin Fugere, this 2,000-acre piece of western North Dakota, bought by Kevin's grandfather in 1930, is the only place he has ever called home. The Fugere's land borders the Little Missouri National Grassland: nearly 800,000 acres of topographical and visual relief from the surrounding flatness of the Great Plains. Thwarting the orderly compulsions of the highway engineers, none of the roads runs straight in badlands of 3,000-foot buttes, rugged draws, and the meandering Little Missouri, the river that defines the area.

I parked beside a red barn. Scattered around were a couple of snow-mobiles, a tractor, a tire swing hanging from a tree, a series of corrals, some sheds, and to the south and west some hills. At first I didn't see anyone, but soon a distant figure on horseback appeared off to the west. It was Kevin Fugere, driving a couple of dozen head of cattle, bringing in some of his stock for branding. When Kevin dismounted, I saw that he was about 40, wore glasses on the bridge of his hawk nose, kept the hair under his cap pretty short, and stood maybe a boot heel over six feet. As Kevin and I spoke, the rest of the branding party trickled in: Kevin's sister Lori and her husband, Dwayne Shypkoski, who own a nearby ranch; their two kids, Trell, a six-year-old boy, and Shea, a four-year-old girl; and Kevin and Cindie's four-year-old son, Ian. About midway through the branding, Cindie returned home from her job with the North Dakota Department of Agriculture.

One by one, calves were dragged from a holding pen in the corral. Kevin and Dwayne toppled each young cow to the ground, pinning it tightly while Lori applied the branding iron. If the calf was male, Kevin made it into a steer with a few deft strokes of his knife. Shea gathered the shorn testicles in a bucket of cold water, destined to become what westerners euphemistically call Rocky Mountain oysters. Trell knelt beside the bawling calf and sprayed the cut with antiseptic. Some folks, especially urban dwellers, may feel squeamish about branding and castration (to watch Lori artificially inseminate a cow would have shocked them even more), but to the Fugere clan, such doings are just another part of daily ranch life. Everyone pitches in, the kids learn ranch ways from an early age, and time flows on, one generation passing into the next. That night, after a hearty dinner at the Fugere table, I walked into another room in the old two-story house. There I found one of Ian's teddy bears lying on the floor. Blue and red markings colored its otherwise white hair. Puzzled, I picked up the bear and asked Cindie what had happened. She laughed and said the kids had branded it with felt markers.

Like many ranchers, the Fugeres love the ranching life. They worry about threats that could diminish or even end that life—competition from the global marketplace, the oligopoly that runs the packing industry, drops in beef consumption. The hazards are many. But one of the biggest threats is a pretty plant all gussied-up with heart-shaped bracts resembling greenish yellow flowers, a species that many veteran grassland weed-whackers dread most of all: leafy spurge.

Leafy spurge is no newcomer. This Eurasian species first showed up in 1827 in Massachusetts, perhaps brought over accidentally or as an ornamental. A hillside of those distinctive greenish yellow bracts waving in the breeze does look quite fetching; at least, it does to an unknowing observer who isn't biased by fear and loathing.

Euphorbia esula heeded Horace Greeley's advice and went west with the pioneers, both spreading on its own and jumping ahead via contaminated grain shipments. It reached North Dakota in 1909. By the 1990s, leafy spurge also had reached 36 other states, infesting from three to five million acres altogether. The northern Great Plains have been hit hard, North Dakota the hardest of all. And unless it is controlled, leafy spurge has a lot of ground left to cover. Calvin Messersmith, a leafy-spurge expert at North Dakota State University, notes that *Euphorbia esula* can grow on almost any terrain, in environments from dry to subhumid and from subtropic to subarctic. He says that after leafy spurge is established in a new area, there doesn't seem to be any topographic limits to stop it from invading adjacent land.

Unlike the explosion of cheatgrass or yellow starthistle, the spread of leafy spurge through the Great Plains isn't tied so closely to disturbances caused by livestock. The grasslands of the plains differ fundamentally from the grasslands of the intermountain West. Remember that bunchgrasses and a cryptogamic crust typically covered the ground in the prelivestock intermountain West. But the native grasses of the Great Plains are rhizomatous, meaning they send out more or less horizontal underground stems to form a thick mat of roots and stems. This is the sod that was used by many pioneers to build their prairie houses. Above ground on the pristine plains, one would have encountered an unbroken sea of grass, not the sporadic tufts of bunchgrasses found west of the Rockies.

Scientists think that rhizomatous grasses thrive in the plains in part because the rains largely fall in the spring and summer. This differs markedly from the pattern of the intermountain West, where precipitation mainly occurs in the fall and winter and drought rules the summer. As a result, western bunchgrasses go dormant during the summer while the grasses of the plains stay photosynthetically active, or green, throughout the hot months. Richard Mack, a professor of botany at Washington State University and a leading expert on grasslands, thinks the summer dormancy of bunchgrasses may account for the fact that bison always have been scarce in the intermountain West.

To maintain a supply of milk, bison mothers need green forage for several months after they bear their calves in the spring. No summer grass, no calves; no calves, no bison.

Because the summer rains fall mainly on the plains, most of the 30 to 70 million bison that once inhabited North America roamed the Great Plains. Mack explains the dynamic relationship between bison and plains grasses: "Bison cannot be likened simply to large lawn mowers; these animals have additional impact through their trampling, wallowing, and defecating. Trampling may be especially important, as is indicated by the very different ways plants may respond to it. If a piece is kicked away from the mat of a rhizomatous grass, for example, it may be replaced by propagation from the sides of the hole; the dislodged piece may even reroot. On the Great Plains," Mack continued, "bison have long been agents of natural selection, encouraging the evolution of plant features such the production of prostrate rhizomes, that enable the grasses to tolerate the animals' activities. And as is characteristic of coevolution, cause and effect have looped back on each other: through their disturbance of these grass-lands, the bison selected for features in the grasses; the summer-active grasses that persisted then maintained the bison, ensuring further disturbance." Thus bison pressed the plants of the Great Plains to evolve defenses that make them less vulnerable to cattle, though over-grazing still can and does occur and partly accounts for the spread of leafy spurge. But leafy spurge has its own ways, too, and can force its way into areas in which there has been little or no disturbance.

Leafy spurge is a perennial that grows to heights from one to three feet, rising on a central stem from which long, narrow leaves splay out. Its natural advantages over most native grassland plants make it a Rottweiler among dachshunds—that is, devilishly hard to control. For example, the underground shoots and the roots bristle with vegetative buds; one study counted from 35 to 272 buds per root. A researcher once held up a two-inch piece of leafy spurge and told me that a new plant could grow back from the bud on that little fragment. And killing these buds is extremely tough, some lie as much as 30 feet underground. Then, there's also the seeds, which are particularly troublesome. Each flowering stem of leafy spurge produces about 140 seeds in lobed capsules that hold three seeds each. Many people think birds eat spurge seeds and spread them a great distance, but when a leafy-spurge stem matures, the capsules explode and hurl the seeds from one to 15 feet, distributing them evenly and widely. The seeds are also notorious swimmers, floating down water courses to new territory. And spurge seeds remain viable in the soil for eight to ten years.

Then there are those diabolical roots. They grow fast: in four months, a seedling can sprawl roots three feet down and three feet out. They grow deep, usually down to the water table: the main vertical roots thrust their grip to depths of more than 25 feet. They grow horizontally: the parent plant fans out roots in all directions as rapidly as 15 feet a year. They feature vegetative buds: new shoots can grow from these buds, helping spurge spread or recover from damage. Scientist R. T. Coupland and colleagues conducted an experiment in which a leafy spurge patch was excavated to a depth of three feet, then the crater was packed with clean soil. Within only a year, spurge shoots appeared aboveground. For five years, new shoots kept cropping up. Small wonder that ranchers and land managers have not yet controlled this tough invasive species.

Ranchers keep looking for ways to kill leafy spurge because cattle develop an aversion to it. It's not about taste; leafy spurge sickens them. When the plant's skin is broken, it bleeds a sticky, white latex that irritates the stomachs of cattle, produces lesions around their mouths and eyes, and causes weakness and scours. Just walking across the freshly mowed stubble of leafy spurge can cause dermatitis on the feet of horses. The latex can produce dermatitis in humans, too, and it stings badly if you get it in your eyes.

The nasty latex also alters cattle behavior: They learn to avoid leafy spurge. Ranchers long have known that cattle shun or underutilize pastures laced with leafy spurge. Research indicates that cattle eat less than half the available forage on a range even when leafy spurge constitutes a mere 10 percent of the plant community. When cows eat less forage, ranchers make less money; they simply can't raise as many head of cattle. In Montana, researchers found that rangeland capable of supporting 20,000 cows could support only 6,600 once it had been infested by leafy spurge. In 1990, a 1,200-acre Oregon ranch sold for only 10 percent of the value of comparable properties because leafy spurge had rendered it almost worthless.

North Dakota ranchers are hurting worst of all. A study estimated that ranchers in that state lose about $24 million a year to spurge. Furthermore, as in a natural ecosystem, a ripple effect from spurge losses spreads out through North Dakota's economy. The same study found secondary impacts of about $54 million. Add in some smaller costs, and the annual total for the state is about $86 million. That's the reason that Cindie Fugere lives and breathes leafy spurge when she's away from the ranch, though she's starting to diversify to other weeds, too. Since 1992, she has been the noxious-weeds coordinator for the North Dakota Department of Agriculture.

On a sunny June morning Cindie Fugere and I drove into the stark beauty of the badlands to visit people involved in anti-spurge projects. She is in her late 30s, tall, slim, with short brown hair and bright blue eyes. Somehow Cindie blends intensity with affability. When it comes to leafy spurge, that intensity often surfaces. She wants action. She wants people, especially those higher up the political food chain, such as the feds, to "get out of the sand box of research and science and implement the technology." She gets impatient because she knows that the weed isn't waiting around while people study and plan. Fugere notes that many local towns and counties are actively fighting leafy spurge. They allocate significant amounts of public money for control, money that could be going to schools. She cited the case of Amidon County. Though it's a poor area, the members of the grazing association voluntarily tax themselves 30 percent extra to fend off spurge. They don't have it bad yet, and they don't ever want to have it bad. "Leafy spurge is a common subject in coffee shops," said Fugere. "It's a part of life here."

Out in the badlands, Fugere and I pulled in beside the pickup of Tom Bakken, who is one of maybe half a dozen ranchers working together on the Deep Creek Weed Innovation Network (WIN) Project, one of the programs Fugere oversees. Bakken came over to shake hands. Not exactly a standard-issue Dakotas rancher in appearance, Bakken sported a full red pirate's beard—with bandanna, earring, and ponytail to match. He drove us up a jolting dirt track in his thoroughly dented pickup, complete with cracked windows and a door that opened only when you put your shoulder to it. A burly man who probably could cut a wide swath with a cutlass, Bakken instead gripped a shovel in his hand. Were we going to dig for buried treasure? In a sense, yes.

After we parked and walked up the hillside to a patch of leafy spurge, Bakken booted the shovel into the earth at the base of a sickly looking leafy-spurge plant and flipped a few loads of soil onto the ground. He and Fugere kneeled down and sifted through the soil with their hands until they found their treasure: *Aphthona*, or flea beetles.

Ranchers call them bugs. Scientists call them biological-control organisms and refer to their use as biocontrol. Hundreds of species of foreign insects have been recruited to help fight the war against invasive plants and animals in the United States. Researchers brought over several species of this particular genus of flea beetles from leafy spurge's homeland to do battle with *Euphorbia esula*. Because *Aphthona* coevolved with leafy spurge, it's generally more effective than our native insects in knocking it back. These little fellows accounted for the sickly look of that leafy-spurge plant. They bore into the roots and feast on the innards. Bakken says the bugs seem to have kept the leafy spurge down a bit, though they've only been around for one year. Researchers say it usually takes three to five years before the flea beetles get into high gear.

Bakken hopes that the bugs and other innovative approaches to leafy-spurge control will lessen his reliance on chemicals. He has been spraying herbicides on leafy spurge for 25 years, and all he's gotten is expensive, intermittent relief—and a sore back from lugging backpack sprayers. "Spraying takes a lot of time," said Bakken. "I spend a hundred hours or better just spraying every year." But he realizes that he must fight leafy spurge. "You know if you don't do something that it slowly takes over the grass, and as it takes more away from the grass, you end up having to put less cattle on it." In addition to the bugs, Bakken likes the coordination of the WIN project. "It used to be that the guy up the crick from you didn't [control leafy spurge], and all the seeds came down and you were back at it again. It's nice in this project to get neighbors together doing the same thing." Indeed many people are joining together. Fugere and company have sown WIN projects around the state, and bugs have been planted on thousands of sites. Fugere harbors even bigger ambitions: she wants long-distance neighbors—people throughout the drainage of the 456-mile-long Little Missouri—to work together to control leafy spurge.

A few miles down Deep Creek, Fugere and I stopped at the main compound of the HT Ranch, another member of the WIN group. This historic spread now encompasses 8,400 acres, but in the open-range era of the 1800s, the HT empire covered hundreds of thousands of acres. Shortly after we arrived, a group of children showed up to tour the grand, 19th-century ranch house, now a museum piece.

The current owner, Paul Herauf, met us at the ranch compound. His grandfather bought the HT around 1940. Just shy of 40 years old, Herauf said, "I've been fighting spurge for as long as I can remember." Like all ranchers since the chemical-pesticide boom of the 1940s, the Heraufs automatically turned to herbicides when leafy spurge invaded, but they've had no better success than Bakken has. "I don't think you could use herbicide to actually get spurge completely under control," said Herauf. "And it's hard on other things, too. If you hit a tree branch or something [with pesticides], it kills it right now." He added that herbicides can be a pain because they can't be used when it's too windy or too hot or too near water. Finally, like all ranchers, Herauf laments the effort and expense of chemicals.

Herauf also runs bugs on his land, and he has high hopes for them. But on the day we visited, he was eager to show us another weapon in the anti-spurge arsenal. He folded his tall frame into his truck and drove Fugere and me across his pastures to the section of Deep Creek where we could watch the goats at work. Goats, living up to their reputation as indiscriminate eaters with iron stomachs, find leafy spurge palatable. In fact, they actively seek it out, especially in the spring and early summer when it's young and succulent. As we walked down to the bank of the creek, where the spurge was worst, we saw the first of the goats come around a bend about 200 yards away. They moved at a beachcombing pace, slowly strolling, stopping frequently to nibble. When the leaders reached us, the rearguard was still coming around the bend. This was a big herd of some 1,100 angora goats. Herauf didn't own this four-footed spurge squad, however. They belonged to a fellow who loaned them to the WIN project in return for the free forage. The herd ranged up and down the 15-mile length of Deep Creek, grazing leafy spurge on all the WIN members' ranches. The goats were kept in line by herders, one of whom we saw riding horseback along the outside flank of the nibbling mass. The herder and his darting canine shepherd had to work to confine their individualistic charges to a narrow strip along the creek, where leafy spurge was dense. The herder and friend also kept the goats moving; otherwise they would eat other plants when they'd finished the leafy spurge.

We stood still, rocks in a stream, as the goats flowed around us. I saw that they did indeed select leafy spurge most of the time, chomping the leaves and top growth and leaving just a scarred stem. Given the root system, this didn't kill the leafy spurge, but the goats sure set it back some and kept it from producing seeds. Herauf believes the trio of goats, chemicals, and bugs could keep leafy spurge at bay.

From the HT we headed north, slaloming past the buttes of the national grasslands. Fugere talked about her personal struggle against leafy spurge. She and Kevin constantly patrol their home ranch, and so far have kept it free of spurge; but they also just started running cattle on a chunk of federal land of which about 30 percent is heavily infested. Fugere reported that the 170 head they put on that land the first year came off a hundred pounds lighter on average than cattle grazed on spurge-free pasture. She added that going into the winter a hundred pounds light is dangerous for a cow.

Badly spurged land sometimes leads to overgrazing, even in the cattle-resistant Great Plains. Ranchers are accustomed to putting a certain number of cows on a certain parcel of land; they count on it. When leafy spurge reduces the useful size of that parcel, ranchers often continue to run the same number of cattle on it, which causes overgrazing. This pattern can degenerate into a downward spiral. Overgrazing promotes the spread of leafy spurge, which further shrinks the percentage of the parcel that cows will graze, which in turn leads to even worse overgrazing (assuming the rancher continues to run the same number of cattle), which allows leafy spurge to spread farther, and so on until the parcel is lost.

As Fugere and I continued through the land she loves—past the striking rock formations, the grassy hills, the verdant bottomlands of the Little Missouri—our conversation, in sync with the landscape, shifted to the big picture. "There's no money compared to the kind of work that needs to be done to improve the range," she said. She thinks that the government is beginning the hard work of protecting and restoring the land, but too often, she feels, government doesn't follow through. "In every case in agriculture, the health of the land is about someone going out and doing some physical labor. It may take a lot of people to do weeds, and federal land managers don't have a lot of people. It may take a lot of herbicides, and the federal land-management system still has a lot of barriers built around restricted-use herbicides. So even though we're watching the system change, watching people set new priorities, what you see happening frustrates you because you know you're still so far from the real goal."

About 20 or 30 miles down the road, the badlands got less bad and soon the landscape settled down into rolling hills covered with grass— and leafy spurge. We turned into a ranch yard squirming with puppies from a recent litter. Amid them stood Pete Wirtzfeld, a young, third-generation rancher. The Wirtzfeld clan has employed a variation on the goat theme to combat leafy spurge. They have made sheep their tool of choice, though some members of the ranching community condescendingly refer to sheep as range maggots. For decades, Wirtzfeld's grandfather and father raised cattle while battling leafy spurge with herbicides, but the weed continued to plague them. Finally about 30 years ago, they decided to give sheep a try, and they've been raising them ever since. "A lot of these dyed-in-the-wool ranchers thought sheep'd graze grass to the ground," said Wirtzfeld, "but it's just a matter of management."

We stood with Wirtzfeld and looked down a road that separates his ranch from a neighbor's land. The contrast brought to mind a paint-by-numbers scene, with the light green of healthy pastures on Wirtzfeld's side of the road and the yellow-green of leafy spurge on his neighbor's side. Wirtzfeld pointed out some places where grass still grew amid the spurge, though little would survive. He said that after the neighbor's cattle had badly overgrazed the spurge-free parts of the ranch, the desperate cows still wouldn't enter the leafy-spurge areas to get at that remaining grass.

To show us a different picture, Wirtzfeld took us down the road to the border with a different neighbor. Wirtzfeld stopped across the fence from some sprawling bottomlands that appeared to be in good condition. "Leafy spurge used to be so thick down there that when you'd drive a three-wheeler through it, the front wheel'd ride up on the spurge, lift clean off the ground onto the thick mat of spurge." Wirtzfeld's success with sheep finally persuaded this neighbor to try sheep; according to Wirtzfeld, in two or three years the leafy spurge had been suppressed in the bottomlands.

As effective as sheep and goats can be, one shouldn't mistake them for a silver bullet that will cure what ails leafy-spurge country. As Wirtzfeld notes, they do require more management, more herding. They also can cause problems, depending on the site. For example, goats will defoliate desirable woody plants. Market considerations also cloud the bright picture: there is little demand for goat products. However, Fugere does think the wool market is strong and that many ranchers in leafy-spurge territory would do well to raise both sheep

93

and cattle. It's the kind of new approach that the weed invasion requires. But most ranchers still keep sheep at arm's length, largely out of cultural inertia. Fugere thinks traces remain of the nineteenth century's range-war mentality, when cattlemen and sheep herders were at each others' throats. Smiling, Fugere said that she even sees some resistance in her own home, noting that Kevin has trouble getting excited about using sheep or goats on that leafy spurge-infested federal land they lease. More seriously, Fugere added that she understands the immense practical roadblocks to bringing sheep into the cattleman's fold. She pointed out that Kevin works 7,000 acres essentially by himself, that his hours already are very long, and that adding another type of livestock would impose a huge new burden.

Before we left Wirtzfeld's place, he took us out to see his bugs—by now I was convinced that biocontrol agents crawled on every ranch in leafy-spurge country. Dubious of biocontrol, he hadn't visited his flea beetles for a spell. When we saw circles of blighted leafy spurge around the release sites, Wirtzfeld was impressed: "I'm a lot less skeptical than I was five minutes ago." He pulled the top off a leafy-spurge plant and marveled at the withered leaves, its stunted growth, and the fact that it hadn't flowered yet. Suddenly enthusiastic, he began peppering Fugere with questions about the bugs. As he drove us back to our car, he mused about the complexities of managing sheep and bugs together. He figured it required thoughtful timing, that if he first grazed leafy spurge with the sheep, then the bugs would thrive on the new growth. "Producers need to learn about bugs," he said, "about their life cycles and such. You need to know as much about them as you do about cows and sheep." I wonder what Wirtzfeld's grandfather would have thought of wrangling beetles.

Jerry Asher is conducting a slide show. He shows Clarence Seeley, the scientist from the University of Idaho, Moscow, who made the first discovery of yellow starthistle in Idaho. Asher recently photographed Seeley standing at the very spot where he stopped to look at this unfamiliar plant back in 1938, the spot from which yellow star fanned out to infest hundreds of thousands of acres. Asher then shows a slide of rush skeleton weed growing waist high in the Boise National Forest. Seeley first noticed this species, too. In 1954, he came across two skeleton-weed plants along Idaho's Payette River. By 1964, it had expanded to about 40 acres; in 1997, the infestation had spread to about four million acres. Asher shows spotted knapweed in a wilderness area on the Selway River in Idaho, a place where imperiled salmon are further imperiled by erosion from the banks, which knapweed can't bind nearly as well as native plants did. Sometimes he cites a study that shows increases of 192 percent in eroding sediments following takeover by spotted knapweed. Asher also shows slides of squarrose knapweed and perennial pepperweed and leafy spurge and medusa-head and diffuse knapweed and Dalmatian toadflax and whitetop and Scotch thistle and Russian knapweed and tamarisk and Scotch broom, and he hits his stride like a low-key gospel preacher, if there is such a thing, and tells you that many weeds are reaching the upswing in their exponential growth curves; that they're galloping across the West at the rate of 4,600 acres a day, and that only includes federal lands, that the fires are greatly increasing the spread of weeds, that it's an explosion in slow motion and a biological emergency, and weeds are growing out of control, and we gotta do something about it. . . .

Time for a deep breath.

As the national weed outreach leader for the BLM, Jerry Asher has for years travelled the West gathering information about weeds. He fervently promotes an antiweed strategy that generally gets overlooked: prevention. His strategy blends the thinking of the hundreds of scientists, land managers, conservationists, ranchers, and other weed fighters with whom he has met. Asher often uses slide shows in his impassioned campaign. His road show has opened people's eyes to the enormity of the weed invasion. Some members of his audiences get pumped up and are ready to wade into the nearest patch of skeleton weed with a machete. Others just want to escape this bludgeoning and go home to watch TV. But neither reaction is what Asher is looking for (though he'd much prefer the former to the latter). He wants people to get ahead of that exponential growth curve. Asher likes to show a slide of a sign that stands along a road in the Okanogan Valley, Washington. The sign features drawings of spotted and diffuse knapweed and reads, "Control Noxious Weeds. Its Your Responsibility." His point is not that this sign contains a grammatical goof, but that the sign was put up way too late. In 1955, long before someone forgot the apostrophe in "It's," an extension agent in the valley asked for $1,500 to get rid of diffuse knapweed. "He didn't get it because diffuse knapweed wasn't a big problem," announces Asher. "Now it would cost millions and millions, if you could get rid of it at all." Sadly, such penny-wise and pound-foolish thinking still happens in too many cases.

"When concern rises, the battle is lost." Asher likes to repeat this observation made by Tommy Gooch, a BLM range conservationist. Asher also quotes anti-exotics activist Faith Campbell, who said: "While the weed is just showing up, you can't persuade the responsible authorities that it's dangerous. 'We haven't seen it do anything,' they say. And then when it spreads all over the place, they say, 'This is too big; we can't deal with it.'" If authorities do try to deal with massive weed strongholds, they use up their scarce allocation and, in many cases, accomplish relatively little. "A lot of places," said Asher, "must be considered permanently infested until our technology or economics changes." Asher thinks we should spend at least as much time and money on stopping the expansion of weeds: "Let's focus on what isn't infested and keep it that way."

To motivate people to hop on the prevention bandwagon, Asher pulls an ace out of his sleeve. He shows a slide of a healthy native bunchgrass community and announces that only about 5 percent of the nation's hundreds of millions of grassland acres currently is significantly infested. (This percentage discounts the presence of cheatgrass and of minor weeds.) Asher explains that we can protect these lands if we act quickly and effectively. Once they're safe, we can continue with the long-term task of controlling invasive plants in the badly degraded areas. Of course, with enough resources, both could be done simultaneously.

Asher wants the nation to fight weeds the way it fights fires: "We need to do a lot of survey and mapping," said Asher. "If you're going to fight the enemy, you must know where it is. Usually," he added, "we find these infestations accidentally as we're conducting other activities. That needs to be replaced by a systematic detection process." Right now when one hears that four million acres of rush skeleton weed infest Idaho, that's just a rough—though educated—estimate. Better information also would answer the question, "How extensive is the infestation?" Experts reckon that weed densities run the gamut from a few plants per acre to monocultures, but there's little data. "It's sad that the agencies don't know these things," said Asher.

After detecting where weed outbreaks have occurred, then comes the initial attack: get the weeds fast—before they spread. Asher cites an exemplary case in the Pueblo Mountains, out in the remote reaches of southeastern Oregon. A BLM employee noticed a small patch of yellow starthistle back in 1992. He pulled every one of the 700 plants. Knowing the wily ways of starthistle seeds, he returned the following year with a colleague and found 2,000 plants, all of which they yanked. He continued watching over the site, pulling 6 plants in 1994, 30 in 1995, and 6 in 1996. He'll keep going back until his counts hit zero for a while. "So even though there are a million and a half acres of susceptible land in that area," Asher proudly reports, "those people are not going to let yellow starthistle become a problem."

That BLM employee in the Pueblos demonstrated the most efficient way to design a systematic prevention and detection process: use people who already travel the grasslands for work or pleasure. The list of potential recruits includes road maintenance crews, ranchers, hikers, mining employees, boaters, firefighters, anglers, and campers as well as the people throughout the federal and state land-management agencies. Tina Ayers, a botanist at Northern Arizona University, takes local native-plant society members down the Colorado River to locate and remove ravina grass from beaches in the Grand Canyon. Verne Wright and family, who own a ranch in Idaho, ride horseback on both their land and nearby federal lands with backpack sprayers at the ready. In Stehekin, Washington, a tiny town in the remote mountains of the northern Cascades, people are highly motivated to seek and destroy weeds; for each bag of spotted knapweed they bring in, they get a bag of doughnuts.

Watchfulness—another of Asher's goals. His optimism really gets cooking when he speaks of people being aware of weeds the way they're aware of litter. He imagines spinoffs from the Adopt-a-Highway program. Rick Wallen, when he was the weed coordinator at Grand Teton National Park, set the standard for alertness. A few years back, a forest fire broke out in the park and a firefighting camp was set up near a thick patch of spotted knapweed. Before the firefighters could tramp through the knapweed and inadvertently carry it to untainted areas of the park, Wallen rushed out and cordoned off the knapweed site.

Simultaneous with detection and eradication comes nursing susceptible lands back to health so they're less vulnerable. Land managers should take heed of the Hippocratic Oath—first do no harm. "It all comes down to three words," said Asher, "Disturbance, disturbance, disturbance. Disturbance from grazing, from off-road vehicles, from mining, from old homesteads, from logging roads . . . and not much land out there hasn't been disturbed." Though farming eliminated much of the nation's grassland, grazing is by far the biggest disturbance on what remains. "Grazing should be a process to improve native-plant communities, to improve the health of the land," said Asher. "To avoid weeds is to avoid overgrazing."

At one point during Asher's slide show, he projects an image of a burn in a ponderosa-pine community in the Selway/Bitterroot Wilderness in Idaho, one of the wildest places left in the lower 48. The district ranger was pointing out the rapidity with which spotted knapweed expanded following a fire. Asher told the ranger's sorry tale, how for a couple of years following the fire, he wanted desperately to send crews into that burn site to destroy the knapweed seedlings before they could become established. It would cost only a fraction of what it cost to put those fires out, but the ranger didn't have the authorization to spend money on such work. "It's an example of what a policy change could do," said Asher.

That slide also shows what a little funding could do to combat weeds. Asher and a group of other land managers and scientists got together a few years back to roughly calculate just how much money it would take to implement a lean but effective program in the West to reduce the spread of weeds by 90 percent on BLM lands, which comprise some 180 million acres. They came up with a figure of about $10 million. More officially, in 1997, BLM developed a detailed implementation plan for their lands that not only would carry out Asher's containment strategy but would also restore areas highly susceptible to weed invasion and address, though hardly cure, the big infestations. Their price tag? $25 million. These amounts are chump change in the context of a $1.7-trillion federal budget, even in an era of government cutbacks. If that $10 million or $25 million got the job even half done, it would be taxpayer money well spent.

Support in the battle against invasive weeds may get a boost. A national effort to control invasive exotic weeds of all sorts may come from FICMNEW (the Federal Interagency Committee for Management of Noxious and Exotic Weeds, pronounced "fick'-muh-new"). FICMNEW represents a much needed federal effort to tackle invasive plants in a serious, organized way. According to Deborah Hayes, a U.S. Forest Service staffer and the committee's co-chair, there has been increasing concern about weeds at the grass-roots level, at the trade-group level, and in government. FICMNEW's member agencies are starting to make changes to address noxious weeds.

Hayes acknowledges that the federal government has lagged on the issue of invasive plants. She attributes this in part to the absence of a solid dollars-and-cents appeal. "We're trying to build an economic case to hit weeds early and prevent big problems," said Hayes. "Everyone knows that [invasive plants] cost us money, and that it's better to address it now than later, but the information hasn't been brought together in a cohesive argument, and we're trying to do that now."

Hayes thinks that the inconspicuous and complicated nature of the problem also makes invasive plants a tough sell. "It's easier to just pay attention to what you've got to do tomorrow, and to think about weeds as something to do the day after," she said. "We know ecologically that it's fire coming down at you, but it's hard to visualize it. It's more difficult to communicate the problem to the public than something like Smokey the Bear or 'Give a Hoot, Don't Pollute.' When the public views a landscape and sees green, they think that's good. Weeds can look like pretty flowers. It takes a higher awareness level to understand the impact of weeds than it does to keep trash out of the woods or to put out a campfire."

In his quest to make the dangers of invasive plants clear to the public and policy makers, Asher developed a succinct statement. To be sure he wasn't exaggerating, he ran it by a host of resource professionals and weed scientists. Most found it accurate; those who disagreed thought it too understated. Though Asher refers to his region of concern, his words apply to places all around the nation where exotic plants are invading. "Thousands of watersheds in the West are rapidly undergoing the greatest permanent (given today's economics and technology) land degradation in their recorded history," Asher writes. "This constitutes a state of biological emergency." Asher worries about sounding like an alarmist, but is it alarmist to sound the siren when the building is on fire?

CHAPTER 6

THE SILVER CLOUDS
OF CALIFORNIA

Clouds became an everyday part of Imperial Valley life in 1991. But these weren't the life-giving clouds that valley farmers welcome.

"You can see a cloud from a quarter mile away because they're so dense. Some clouds may be but five or ten feet across, some might be 50 feet across." Farmer Larry Gilbert laughed ruefully and added, "That's a lot of whiteflies. . . . For some reason, especially when whiteflies go over a dry field, they'll clump together. It wasn't a question of if something was there, because they were just all over, almost like a swarm of gnats."

"You'd walk through the fields and you'd get huge numbers of white-flies," recalls Nick Toscano, a whitefly expert from the entomology department of the University of California, Riverside. "You'd sit down to look and they'd get in your eyes. You'd run and they'd get in your nose. There were just that many." Toscano said people began wearing surgical masks. The numbers grew so enormous that the silverleaf whiteflies were killing plants just by sucking them dry. "We'd see nymphs on a plant, and I'm not exaggerating, they looked like they were stacked three high. When they used their feeding stylets [the strawlike body part that penetrates and sucks a plant], they had to wind it through other nymphs to reach the leaf."

Steve Finnell also remembers. Finnell works as a pest-control advisor (PCA) for eight or ten growers in the valley. In order to advise his clients, he goes out on reconnaissance missions; it's especially important to know the enemy's numbers before suggesting remedies. Finnell recalls that his usual careful examination of possible host plants proved unnecessary with whiteflies. "I'd just walk over to a plant and kick it. The whiteflies would just explode." Finnell said that matters went from bad to worse when the whiteflies got airborne. "If the wind was more than one or two miles an hour, you'd have to walk like this"—he hunched over, tucking his head, and stepped backward—"if you didn't, you'd be swallowing whiteflies."

Though weak fliers, whiteflies do labor into the air and drift on the wind when they need to find more plants to feed on. These drifting masses became so common, particularly during the summer, that in 1991 Imperial Valley residents almost saw the whitefly clouds as a threatening weather phenomenon.

Larry Gilbert has dark hair, a lean build, and a narrow, craggy Lincolnesque face. He was born in California's Imperial Valley and has run a farm there for more than 35 years. In 1991, he was farming about 700 acres, some of it in cotton. Early in the year, whiteflies hadn't touched Gilbert's fields. Then he noticed a pattern: whiteflies started building up, notably on spring melons, and then moved over en masse onto cotton and other summer crops. "We hadn't been having any problems," explained Gilbert, "and then along about August my entomologist said, 'You know, we've got to spray this for whiteflies.' So we sprayed it, but three or four days later it looked like there were just as many as there had been. The whitefly defoliated that crop before we were ready to pick it. Wiped it out."

Silverleaf whiteflies harm crops in ways other than sucking them to death—though often they suck them only half to death, which can be enough to make produce unmarketable. Whiteflies also inject salivary fluids into their hosts. For example, their spit leads to irregular ripening of tomatoes, which has inflicted major losses on growers in Florida. Brimming with even more unsavory habits, whiteflies can damage some crops by drenching them with what is euphemistically called honeydew: an excretion of partially digested sap and leaf sugars. "Honeydew on cotton is very serious," explains Eric Natwick, University of California Cooperative Extension Entomology Farm Advisor. "When the cotton goes to the mill, any stickiness can . . . end up shutting down the whole operation, which costs them thousands of dollars an hour." As an agricultural advisor in the Imperial Valley, Natwick gets some of the first phone calls from farmers when bugs show up. He had received a lot of calls in the fall of 1990. He got a lot more from the Imperial Valley farmers in 1991.

The valley lies on the Mexican border, in the southernmost area of southern California. More significant, it lies about 100 miles east of the southern California metropolitan sprawl. It is decidedly not metropolitan—the largest city in the Imperial Valley is El Centro, with a population of slightly more than 30,000. In 1997, I entered the valley from the west, winding down through dry, rock-pile mountains to the flat valley floor. Though the pass measured 3,241 feet, I descended nearly 3,300 feet before leveling out. Most of the valley lies at an elevation lower than the ocean's, and locals delight in marking sea-level lines high up on water towers, silos, warehouses, and the like.

By nature, the Imperial Valley is creosote-bush, reptile-ridden, burning-sand desert, an offshoot of the Sonoran Desert called the Salton Sea Trough. On the midwinter day when I arrived, temperatures in the almost nonexistent shade climbed into the seventies, and no clouds dimmed the brilliant sunshine. Actually, clouds seldom appear above this parched land, which on average receives 2.4 inches of rain a year. Yet if you travel to the valley, even during the searing summer months, you see hundreds of thousands of acres of radiant foliage as green as a golf course.

Soon after Interstate 8 leveled out, I saw the first of the maze of canals and ditches that creates the luxuriant verdure in this desert. Irrigation water is the mother's milk of the Imperial Valley. Perhaps in response to the political controversies that swirl around water rights in ever thirsty southern California, someone draped a large banner across a stack of hay bales along the highway; it read, "Food Grows Where Water Flows."

Irrigation water first flowed into the valley in 1901, courtesy of Charles Rockwood. The water came via a diversion channel from the Colorado River. The area once had been known as the Valley of the Dead, but Rockwood and his development partners immodestly renamed it "Imperial." Within a year, the arrival of water led to the arrival of a couple of thousand people and many thousands of acres of crops. Not long afterward, the first exotic pests arrived, too. This was pest heaven: warm temperatures year-round; few native competitors; a wide variety of well-watered plants available throughout the year.

Whiteflies were first recorded in the valley in 1928. Though many species of whitefly exist throughout the world, and look virtually identical, they can behave quite differently. The species that appeared in the Imperial Valley in 1928 was the sweetpotato whitefly, *Bemisia tabaci,* which does damage primarily through its formidable ability to transmit plant viruses. Worldwide, viruses spread by the sweetpotato whitefly account for more than 40 crop diseases.

The first significant damage in the valley by sweetpotato whiteflies occurred in the 1950s, when the insects began passing a leaf-crumple virus to cotton, one of the valley's major crops. Problems escalated in the 1970s when these pinhead-size bugs started spreading another virus: squash-leaf curl. And yet another whitefly-carried virus, known as lettuce infectious-yellows, soon entered the fray. But *Bemisia tabaci* hit cotton hardest of all with the "honeydew" excretion they produced, heavily damaging the Imperial Valley's star crop. By 1981, according to Eric Natwick, the valley was a disaster. Natwick, who started his job as the county's extension advisor in 1981, estimates that growers lost more than $100 million to the sweetpotato whitefly and its virus allies. In the mid-1980s, *B. tabaci* finally faded to manageable levels.

Knowing the history of the sweetpotato whitefly, one can appreciate Natwick's uneasy feeling when he began getting farmers' calls in 1990.

"I was called to look at a melon field where the plants were wilting," said Natwick. "The grower said the problem had been caused by the drift of defoliate from a cotton field adjacent to his melons." Natwick recruited his office's pathologist, and they drove out to have a look. "We walked a few rows into the field and I turned a leaf over. The underside was completely white from whitefly adults like I'd never seen before. I've seen high numbers, but never where you couldn't see the surface of the leaf. As we looked at more leaves, the adults flew off, and we found the leaves completely encrusted with whitefly nymphs [insect young that haven't completed metamorphosis]. They were actually overlapping each other." It was at that point that the red flag went up. There was something abnormal about the sheer numbers. Natwick sensed that this was no ordinary pest outbreak.

There was something else that worried Natwick about the whiteflies he first saw in 1990. He got a call from a PCA in the neighboring Coachella Valley who reported that whiteflies were causing red cabbage to wilt. Natwick's worries grew, because sweetpotato whiteflies don't typically molest cabbage. A few days later, some PCAs in the Imperial Valley told him that broccoli and cauliflower were under whitefly attack and wilting. "Then I was certain," said Natwick. "We had a different animal than we'd been dealing with before. During the 80s you really didn't see a problem with whiteflies on these crops."

Natwick figured he was encountering a pest that researchers had found in southern states as early as 1986—what was then thought to be a new strain of sweetpotato whitefly. Judith Brown, a plant virologist at the University of Arizona, began to suspect the presence of something different while she was studying whitefly-transmitted viruses. In an attempt to raise subjects for her experiments, she had tried unsuccessfully to propagate sweetpotato whiteflies on poinsettias. Then in the late 1980s, commercial greenhouse owners in Florida complained that sweetpotato whiteflies were destroying their poinsettias. Brown remembers thinking, "What the hell is going on here?" Her sense that a new whitefly had appeared grew when whiteflies began moving onto tomatoes, a crop sweetpotato whiteflies in the western hemisphere ignore.

After much investigation and debate, scientists decided this whitefly wasn't just a strain of the familiar *B. tabaci*, but they still haven't decided exactly how to classify it. Ray Gill, an insect taxonomist with the California Department of Food and Agriculture and a whitefly specialist, considers the new whitefly to be a separate species, but he readily acknowledges that more research is needed and that his learned opinion is somewhat subjective. Whether this critter is finally labeled a species, a subspecies, a race, or something else, its behavior sets it apart from any other whitefly that has hit the U.S.

Because it turns the leaves of pumpkin and squash plants silvery, the newcomer—the aforementioned cloud-traveler—was named the silverleaf whitefly, *Bemisia argentifolii*. A cloud with a silver lining? Not likely. When the press got wind of this, the alien invader got the sensational nickname "Superbug." The origins of silverleaf whiteflies remain murky, but some researchers suspect that they came from Africa. Their pathway into the United States also is uncertain, though they may have stowed away on imported poinsettias (a bit of irony from the standpoint of those greenhouse owners in Florida). Even if poinsettias aren't responsible for the original introduction, they are strongly implicated in the interstate spread of silverleaf whitefly—just picture all those lovely Christmas plants being shipped across the nation during the holiday season.

Being so small, pests can enter the country and spread via an almost infinite number of avenues. The OTA's 1993 landmark report notes that California border inspectors have found that fire ants enter the country in shipments of carpets, on roofing materials, in empty trucks, in loads of produce, in U-Hauls, and on nursery stock; that European gypsy moths travel on firewood, motor homes, lawn furniture, boats, lawn mowers, and campers; that Argentine ants are transported in the dirt stuck to the mud flaps of trucks; and that in 1986 alone, some 3,000 Japanese beetles were found in cargo planes landing at Ontario, California. An OTA chart listing routes of introduction shows that most exotic pests enter the U.S. as stowaways on plants and in other trade shipments. At least those are the most common pathways known. Actually the chart reveals that we simply don't know how most alien pests get here.

The concept of pest is kin to the concept of weed: both are human constructs. North America teems with native ants, mites, viruses, flies, and hordes of other tiny animals that, should they cross our paths, we would categorize as pests. But these species perform a multitude of vital ecosystem functions, even when they occasionally reach epidemic levels. An outbreak of a native tree-killing bark beetle in a wilderness forest contributes to the dynamic harmony of that community as much as does the killing of an elk by a pack of wolves. Some of the trees slain by the beetles, for example, stand for many years as snags, the preferred home for woodpeckers, owls, and many other species. Beetle explosions also open up the forest floor, creating opportunities for sun-loving plants, and the animals that thrive on them—which in turn sustains the forest's biodiversity. Most people, however, perceive pestilential outbreaks as disasters, much in the same way as they regard floods and fires as destructive. And if that bark beetle in a North American wilderness hailed instead from Chile, alarm would be appropriate. Rather than performing a vital ecosystem function, a non-native beetle more likely would disrupt the community. It could reproduce at harmful rates because as an introduced species, it leaves behind in its native habitat its predators and competitors.

Non-native pests have plagued American agriculture for hundreds of years, and they continue to plague it today. The OTA report estimates that about 40 percent of the most destructive agricultural pests are exotic. Just whisper "Russian wheat aphid" in a rural cafe in Kansas and tanned faces will suddenly turn pale beneath their John Deere caps. Mediterranean fruit fly, beet army worm, European corn borer, brown citrus aphid, Caribbean fruit fly, potato blight fungus, diamondback moth, citrus root weevil, varroa mite, Japanese beetle, citrus leafminer, and untold numbers of other non-native pests give farmers grief. They cost plenty, too. The OTA report estimates that each year farmers spend about two billion dollars on chemicals to fight pests, much of which gets sprayed on non-indigenous pests. The Russian wheat aphid alone has been racking up about $170 million annually in agricultural losses. The alfalfa weevil tops it, accounting for about $500 million a year. But neither has matched the infamous boll weevil. Since 1892, when this insect first infiltrated American cotton fields, it has accounted for agricultural losses of more than $13 billion.

Little pests cause a great deal of damage, far more than pigs, goats, mongooses, and other large invasive animals. Some of the most destructive aren't even visible to the naked eye.

Crop-killing microbes worry farmers all over the world. What they fear is what happened during the early and mid-1990s to Jenny and Delan Perry and other papaya growers on the Big Island of Hawaii. The Perry's 70-acre farm lies about 3 miles from the ocean in a rural area near Kapoho, a tiny town on the eastern side of the Big Island. This area is—or was—the heart of the island's papaya region. To reach their farm, I left the blacktop and crunched over a mile of lava-gravel road, curving past abandoned sugarcane fields, banana trees, and coconut palms.

After I met up with Jenny Perry, we drove out into the fields. Perry is petite, fortyish, and fit, presumably from many years of farmwork. She and her husband bought this farm in 1973 and planted a variety of crops, including papaya in 1980. Conditions proved perfect for papaya, and within a couple of years they had planted papaya trees on most of their acreage. The Perrys and the papayas thrived. But in 1992, the Perrys saw evidence of an exotic microbe: papaya ringspot virus had invaded their fields. By 1994, the virus had erupted into a full-blown epidemic. Within a few years, the local papaya industry had nearly vanished—and so had their livelihood.

Perry stopped the truck and led me out into five acres of papaya trees. The couple had chopped down nearly all of their trees—at first in a desperate effort to save healthy trees, in the end because all their trees had the disease and had to be eliminated. But now university researchers, with the Perry's blessing, have turned these five acres into an experimental plot planted with genetically engineered papaya that may better resist ringspot virus. In order to test their efforts, the scientists had left a few dozen of the Perry's diseased trees around the experimental plantings. Aphids transmit ringspot virus, and the researchers knew the aphids would get around to the experimental trees, too. Many of the new trees look healthy, which provides much of what little hope remains for the Perry's and other local papaya growers.

Jenny Perry stopped at one of the old, diseased trees and picked a shrunken papaya fruit. "You get some real ugly stuff," she said, handing me the wrinkled papaya. "They're malformed. More like rubber, because they don't get the moisture or the nutrients." Wayne Nishijima, a plant pathologist with the extension service of the University of Hawaii, Hanoa, notes that the virus seldom kills papaya trees; like any smart parasite, it doesn't want to eliminate its meal ticket. But according to Nishijima, the virus can render a papaya tree commercially useless in about six months. Ringspot virus retards the growth of leaves and turns them yellow, reducing their ability to make food, which in turn reduces the size of the fruit. The virus also reduces the tree's ability to produce sugars, so the fruit isn't as sweet as it could be.

"Once that virus takes hold in an area," said Margarita Hopkins, an economic development specialist with the county, "if it's not eradicated, you can no longer grow papaya commercially. It's just a matter of time before the new plantings of papaya will be attacked." She advocates draconian measures. "The only way to get rid of it is to eradicate all the papaya plants in the area."

Nishijima seconds the motion. In 1992, soon after ringspot surfaced, he and other researchers urged the industry to eradicate all the papaya trees, infected and uninfected, in the immediate vicinity of the outbreak, which was isolated by about two miles from the main papaya-production areas. The industry balked at this idea and opted to only remove infected trees. "They felt they would have suffered too much," said Nishijima. "They would have had to sacrifice maybe a hundred acres of crop." That's about three or four percent of the total acreage that was planted in papaya. Wasn't that being penny wise and pound foolish? I asked. Nishijima nodded in agreement.

Though Jenny Perry and her husband axed their diseased trees religiously, sometimes felling 200 a week, she understands the reluctance of growers to chop down their trees. She points out that a single good tree produces about 40 pounds of fruit, and that the rule of thumb was to cut down the 30 nearest trees along with the infected plant. With papaya fruit fetching about a dollar a pound wholesale, the researchers were asking growers to destroy as much as $1,200 worth of fruit plus 31 trees each time the virus struck. "That's why a lot of growers decided they weren't going to cut," said Perry. "But when they don't cut, they breed the virus for others."

The government didn't force anyone to cut down their trees, but the state agriculture department did mark infected papaya trees with ribbons. Disgruntled farmers often removed these ribbons on the sly. According to Nishijima, growers sometimes threatened the people doing the marking; on a few occasions they even fired warning shots. Sometimes relations among fellow growers turned ugly, as well. Nishijima suspects that occasionally a farmer beset by ringspot would undermine other growers by slipping virus-bearing plants into their fields, either out of envy or in the name of ruthless competition. "There have been documented cases," said Nishijima, "where people have thrown [plants infected with other diseases] into clean fields." Perry and other growers harbored similar suspicions. When ringspot virus first hit, it spread in patterns that seemed unnatural. "If someone got the virus," said Perry, "they wondered how it got there. Did somebody plant the virus here, or did it come naturally? People wondered. This notion of using exotic viruses for agricultural sabotage even affects international relations. A big flap ensued in 1997 when Cuba accused the U.S. of sending an American plane to spray Cuban soil with *Thrips palmi,* a vegetable-crop pest.

In addition to its obvious effects, the virus frayed the fabric of the community. "We used to go and visit each other's farms real freely," said Perry. "We'd check out what everybody was doing, offer suggestions.

Now nobody wants you on their farm because you might [accidentally] bring the virus. Farmers are getting more uptight about visitors coming or even other farmers coming. You call first and get permission, even with people you know well." That tension has lessened during the last year or two simply because now almost every farm has the virus—some 90 percent of the papaya trees have been ruined.

By nature, Perry is a positive, cheerful soul. Yet the extent of the virus's effect on her family and the papaya-growing community gives even a cheerful soul reason to complain. "Well," she said, "Delan had to get a job in town." Other families also had their lives disrupted, including a farmer who started hauling water for a living and one who gets by as a bulldozer operator. "Several farmers got together," she added, "and opened up a bakery in Hilo. Everyone looks to some kind of supplemental income." It's Jenny Perry's turn next; she's taking a teaching job so she can trade places with her husband for a while, give him a turn on the farm. The virus could change their children's lives, too. "My oldest daughter is at the university," says Perry, "and she's wondering how her university bill is going to be paid now that we don't have much of an income."

Jenny says she and Delan used to employ about ten people year-round; now they employ only two part time. "For our employees it's been difficult," said Jenny. "They had to look elsewhere for work." But with the papaya industry wiped out, much less work is available, especially without moving to a different part of the island. The many businesses that served papaya growers have shriveled as well. She spoke of the shippers that used to handle their papaya crop. "They're just not busy anymore. You can go in there any time and they'll unload you. You used to have to wait in this long line. . . . The farm supply businesses are really hurting. Fertilizer places, some of the equipment places. There are a lot of businesses in Hilo closing up." Perry notes that the papaya virus alone didn't cause the closing of all those businesses. The downturn in the sugar industry and ginger wilt and the anthurium blight—two other exotic diseases that hammered local agriculture—have contributed, too.

Perry says that some of the local papaya growers are hanging on any way they can, hoping that the university research will bear fruit, figuratively and literally. However, a year or two ago, most of the Big Island papaya industry moved to a virus-free region of the island. "They're running away from the virus," Jenny said. During the last couple of years, new papaya farms have blossomed around Hamakua, about 60 miles north of the Perry's farm. It turns out that fleeing the papaya ringspot virus is a venerable, international tradition. In fact, the papaya industry first came to the Big Island after the virus drove Oahu papaya growers out of business in the 1960s. So now the torch has been passed to Hamakua, but it may sputter and die all too soon. The virus already has shown up in the Hamakua area.

Though the silverleaf whitefly's arsenal doesn't include viruses, its weapons have proven more than ample. Whiteflies do much of their damage through sheer numbers. They produce new generations every 16 to 21 days, and they believe in big families. Whitefly expert Nick Toscano calculates that a pair of whiteflies can produce about eight million progeny—the females can even turn out young without mating. Like pigs, leafy spurge, and many other major invasive species, whiteflies reproduce exuberantly.

Anyone who lived in the Imperial Valley in 1991 could vouch for the fecundity of the whitefly. Despite the dense pockets of whitefly infestation that Natwick and others saw in the fall of 1990, nothing disastrous happened right away. Winter provided a reprieve because whiteflies are tropical insects and the cold—even the mild cold of winter nights in the valley—knocks them back. It doesn't knock them out, though, and in spring 1991, their numbers started to grow again. By September the whiteflies had become uncountable. "They hit us bad in the fall of '91," said Steve Birdsall, the Imperial County agricultural commissioner since the mid-1980s. No visitor who spent time in the waiting room outside the commissioner's office would doubt Birdsall's claim: on the wall are framed stories from the November 25, 1991, issues of *Time* and *Newsweek* describing the devastation. "The whitefly wiped out fall melons," continued Birdsall. "Over a period of about a month and half, the whitefly took about $13 million out of the agricultural economy, and that's really, really significant." Birdsall points out that the ripple effect from the loss of fall melons cost the valley more than $100 million. Then the whiteflies clouded over into the winter vegetables. "We produce the overwhelming majority of the winter vegetables consumed in the western United States," said Birdsall. "Broccoli, cauliflower, lettuce, things like that. Our early planting of winter vegetables was wiped out, too." Many consumers who wanted to eat their greens encountered low quality and high prices that winter, if they could find their favorite veggies at all.

"It was really a blow to the psyche of the valley," said Birdsall. "They'd never been hit before like this, never had so much of a problem from a single pest. It was especially frustrating because nothing would kill it. People were pretty depressed. They felt help-less." His voice rising, Birdsall added, "Can you imagine a farmer who's gotten right up to harvest with his melon crop, probably put in anywhere from $1,000 to $2,000 an acre just to get that crop there? You've got, say, an 80-acre field, maybe $150,000 sitting there in that crop, and almost overnight it dies. And there's not a damn thing you can do about it. You're spraying and spraying and spraying, and it's dying and dying and dying."

For the year 1991, whiteflies cost the valley's economy about $250 million, but such broad statistics are mosaics of thousands of pieces of individual hardship. "There was turmoil in the community," said Refugio Gonzalez, the county director for the University of California Cooperative Extension. "Growers weren't buying chemicals, fertilizers, seeds, fuel for their equipment, boxes for packing. All of that was gone, lost. And most of that was bought here. All of these allied indus-tries took it on the chin." Thousands of people lost their jobs; 2,000 in the agricultural sector alone. Many were poor farmworkers or day laborers who crossed over from Mexico. Most had no salable skills or financial cushion to fall back on other than the public safety net.

"All of the social agencies were really impacted," said Gonzalez.

"I remember the director of the coordinated housing authority saying, 'We've got lists that are really long of families who are sleeping out in cars because they're not working; they don't have an income.' We had little kids, going to school, living in cars. It had to be tough on them. Where did they shower? Where did they wash their clothes?"

The whitefly also had a direct effect on what passes for urban in the Imperial Valley. The breezes that carried clouds of whiteflies across the agricultural fields also carried them into the towns. While in El Centro, I dropped by the obvious place in which to hear the town's story: the Central Barber Shop.

"Oh! All of a sudden you'd run into a cloud of 'em, like snow. And you'd better not breathe in too hard, no kidding. It'd screw up your lungs. I ain't kidding you. And if you drove far enough, your windshield would be completely white." That was Mel Valenzuela. Four decades of barbering in El Centro had only strengthened his natural gift of gab. For 36 years he had owned the Central Barber Shop; now his nephew, Dan Estrada, the quiet young man cutting hair at the other barber's chair, owns the shop: "He worries about it and I cut hair," says Mel, laughing.

The guy in Mel's chair getting a few millimeters taken off his crewcut was Al Payne. He looked big, even sitting down. His white hair and deeply furrowed face put him in his 60s or 70s. Retired, he has lived in El Centro since 1952. "The local gardens are going to hell out here," he said, his voice gravelly. Mel chimed in, "Oh, the gardens, yeah. The whitefly killed my rose bushes." He snipped a little more of Al's hoary stubble as he continued: "You know the Bermuda grass? They would even affect that. You'd be cutting your lawn, and you'd see whitefly all over the place." Sitting to the side, waiting his turn with Mel's scissors, was Jerry Nakasawa, a fifty-something farmer who used to live in the Imperial Valley. He delivered a similar report about his new home across the Arizona line in Yuma. "There were flies in the yards, the rose bushes. Now there's stuff you can use, but back in '91, there was nothing. People tried soaps, detergents."

In the barbershop's second chair, Mel's nephew was trimming the neat hair of Pat Apalategui, a local fireman. Pat spoke of several times when he'd gone to the scene of a traffic accident in which motorists collided after being enveloped in a fog of drifting whiteflies. "They said they saw what looked like a cloud of dirt, but it was whitefly." A trim, soft-spoken man in his 30s, Pat seemed to shun embroidered talk. He was careful to point out that such accidents were rare, that whiteflies didn't swarm over the entire town all day. Mel nodded and said, "It wasn't like in *The Birds*," referring to the Hitchcock movie. Mel noted that the whitefly invasion didn't shut down the town or anything. "You'd get used to it, like we get used to the 120 degrees around here. Hey, we're tough over here."

After more conversation and a few more snips—though Al's hair still looked the same to me—Mel added, "I'd have to shake a customer's hair out sometimes because it'd be loaded with whitefly." I was appalled, and the momentary look on my face told all. The guys broke out laughing—except for Mel, who kept a straight face when I asked if he was joking. "Barbers don't joke," he quipped. The guys laughed again. Mel was good. Lest I think the whole whitefly discussion was a fabrication, he hastened to add, "The other parts are true."

If I'd had any doubts about Mel's veracity, Birdsall, the agricultural commissioner, would have cleared them up: "This pest was a serious urban pest. You can drive around town and you won't see many hibiscus plants growing here anymore. The whitefly just wiped out our hibiscus population. There are several ornamental plants that whitefly devastates. It caused hundreds of thousands of dollars of damage to our ornamental crops." Birdsall paused, smiled slightly, then added, "It was kind of interesting, though. The whitefly caused the community to become cohesive against a common enemy. I'd have people stop me in the streets, saying, 'Gee, how can I help? Here's five bucks toward the solution.'"

Solution. The word implies taking care of a problem in some final way. By that definition, no solution to the whitefly invasion was found in 1991, none has yet been found—and none is ever likely to be found:

exotic pests of the whitefly's magnitude virtually never get eradicated. It's more a matter of learning to live with them.

Farmers have long since settled into seemingly endless trench warfare, defending their crops against wave after wave of exotic pests. Growers in the Imperial Valley know this as well as anyone. The onslaught didn't begin with the silverleaf whitefly, nor will it end with it. Typically pests accumulate. Each species ebbs and flows according to environmental conditions and actions taken by growers, but seldom does it vanish. Farmers must deal with a relentlessly growing tangle of pests. Imperial Valley growers reacted to whitefly the way most farmers in the U.S. react to pests: "What is it? And what do we spray it with?" they wanted to know. Pesticides have been mainstream agriculture's knee-jerk reaction to pests since the 1940s. The allure stems from several factors. One is the promise of simplicity: if you've got bugs, spray 'em; better yet, just spray your fields periodically and you won't even get bugs—or so the sales pitch went. Another factor is that pesticides allow growers to take immediate action. When confronted with an outbreak, speed is vital. "Many growers don't think of avoiding a pest problem," said Nick Toscano, the researcher at the University of California, Riverside. "They think of a growing season, of having a crop ready for a hot market." But measures taken to control one pest may affect another, for better or for worse. Toscano believes that farmers and scientists usually focus on the worst problem of the moment and hope pesticide use doesn't incite another pest to run riot—which is what happened when pink bollworm came to the Imperial Valley.

Cotton was the valley's most precious crop. According to Refugio Gonzalez, the county's university extension director, "It used to be called 'King Cotton' around here. We had some of the best yields in the world." By the 1960s, cotton flourished on about 144,000 valley acres. But non-native pests, the cotton bollworm in particular, dethroned the king. The bollworm had been a problem since the beginning of cotton growing in the valley. Then in the late 1960s, pink bollworm hit cotton hard, adding its considerable might to that of the cotton bollworm. Growers sprayed for pink bollworm, but instead of controlling that pest, the chemicals cleared the way for an explosion of spider mites in cotton. In the mid-1970s, the tobacco budworm joined the invasion. Around 1980, the sweetpotato whitefly burst on the scene, spoiling cotton by leaving honeydew on it. By the end of the 1980s, cotton cultivation had shrunk to a mere 10,000 to 15,000 acres.

Around the turn of the decade, cotton growers began feeling hopeful again. Restrictions on the timing of the season had greatly reduced pest numbers on cotton. Birdsall says the county was poised to get back into cotton in a big way, "but then the silverleaf whitefly knocked the hell out of us." Instead of climbing, cotton acreages resumed dropping, leveling off today at about 5,000. Calcot, a cotton cooperative that used to market almost all of the cotton grown in the valley, closed down its sprawling storage and shipping facility just north of the town of Imperial. "We're losing the infrastructure," said Gonzalez. "We've lost the gins. We used to have 25 or 30. Now there's just one." Thinking back over the history of cotton in the valley, Gonzalez added, "The pressures of insects continued to grow, so production costs continued to grow as well." Then, when the silverleaf whitefly invaded, "the bottom fell out."

Eric Natwick points out that the shortened growing season has greatly reduced the impact of pests on cotton, and the hardcore growers who refused to abandon cotton are now doing all right. Still, Natwick and others don't think a wholesale return to King Cotton is in the cards. For one thing, says Natwick, farmers have been attracted to other profitable crops. But even if they wanted to take up cotton again, it would be difficult, especially considering that most of them stopped growing cotton back in the early 1980s due to the pest problems. "People who used to grow cotton no longer have the equipment," said Natwick. "They don't even know how to grow cotton anymore. It's been nearly 20 years." In many cases, says Natwick, it was the parents who had grown cotton, and now the kids run the farm.

The difficulty of reviving cotton illustrates some of the obstacles that often make it hard for farmers to shift from one crop to another, even if a pest invasion makes such a switch desirable. Steve Castle, an entomologist with the U.S. Department of Agriculture, points out that the Imperial Valley, like most farming areas, has developed its characteristic tools, its system, its crops, and its markets. Farmers have devised complex schemes to try to beat the rest of the market by a week or to come in with a crop just after that crop gets scarce elsewhere. To change means new equipment, new knowledge, lost opportunities. "Farmers have a niche that has evolved here, and they have to fulfill that market niche," said Castle, "so cultural measures are really quite difficult. Yes, you could take away the melon crops and plant something else [which would greatly reduce whiteflies], but they're tooled and geared up for planting melons, and that's what this valley has been famous for since the 40s." In the end, whiteflies thrived so outrageously on fall melons that in 1992 growers did shift gears. "They went from between 10 and 12 thousand acres . . . down to 80 acres," said Castle. So growers made the switch, but this drastic action sent shudders through the entire valley. "I think there are cultural solutions to the whitefly problem," said Castle, "but it would require such a complete turnaround in the way this valley does agriculture that it would be a revolution."

"The whole system is resistant to change," said Greg Simmons, a U.S. Department of Agriculture scientist stationed in the Imperial Valley. "It's driven by money and fear." Still, Simmons understands that many growers don't have a financial cushion that would ease the transition to a more balanced control program. "In farming, you've got to make your money right now," he said. "If you lose this year, you're dead next year." Simmons also knows that other methods, such as the biocontrol program he and his colleagues are developing, require growers to acquire new knowledge—and these new methods are harder to implement than are spraying regimes.

Despite their reasons for relying heavily on pesticides, many who rely on them have stifled their tendency to spray anything that moves and are looking for alternatives. As shown with pink bollworm, there can be unexpected repercussions from pesticide use. Additionally, chemicals don't work cheap. A few years ago, the Environmental Protection Agency (EPA) estimated that farmers spend about eight billion dollars a year on pesticides. John Pierre Menvielle gave me an example of the sum that resulted from adding together silverleaf whiteflies and chemicals. Menvielle is a leading grower and an Imperial Valley native whose grandfather began farming in the area in 1907. "We had 400-dollar-an-acre pesticide costs," he said, speaking of his broccoli fields. "In the last year that we grew broccoli, our growing costs were 14 dollars an acre and that year we got back $300, tops. So we took an 1,100-dollar-an-acre loss on roughly 140 acres. Who needs it?" Steve Finnell, Menvielle's PCA, was with us in the field. Showing only a little smile, he said, "I think we're still paying for the pesticides we used in 1991." Menvielle no longer grows broccoli.

The high cost of chemicals has prompted many farmers to abandon prophylactic spraying in favor of waiting until pests reach a certain threshold. "Almost all of our spray jobs are close to emergencies," says farmer Larry Gilbert. "You wait until you're sure the pest is going to become an economic problem before you pull the trigger." For example, he explained that a grower might hold off spraying for whitefly until maybe three quarters of the leaves of a crop sported on average two or three whiteflies. Of course in 1991, when hundreds of whiteflies thronged every leaf of Gilbert's cotton crop, thresholds became irrelevant. But typically Gilbert spots pests before they're doing significant damage and keeps tabs on them, waiting until the last minute to spray. He's always hoping that the presence of the whitefly's natural enemies will keep pest numbers below the threshold, or at least permit a delay in pesticide spraying. There are serious consequences to inadvertently killing exotic pests' natural enemies, such as big-eyed bugs, beetles, pirate bugs, spiders, and wasps. These are the so-called "beneficials"; they have not coevolved with the alien pest species and generally don't prey on them as effectively as would predators from their homelands.

Still, they often help keep potential pests at manageable levels.

Consider the case of fire ants. In 1957, the federal government and several states set out to eradicate two exotic species of fire ants that had become serious pests in the southern states. Airplanes loaded with broad-spectrum pesticides began dumping these chemicals all over the 90 million acres that fire ants infested. These virulent poisons killed a bunch of fire ants all right—and just about everything else, including the predators and competitors that had somewhat checked the advance of fire ants. This profligate use of nonselective pesticides left the countryside wide open for recolonization by fire ants; predictably many survived the chemical strafing, and by 1962 fire ants had expanded their territory to 120 million acres.

Nick Toscano thinks that pesticide use probably contributed to the explosion of the sweetpotato whitefly in the Imperial Valley. Remember that *Bemesia tabaci* showed up in the valley in 1928 but didn't become a monster pest until the 1970s. Toscano thinks it is more than coincidence that the sweetpotato whitefly's rapid ascent to major pest status occurred as growers began using pyrethroids to control other pests. The pyrethroids massacred many beneficial species, including those that kept the sweetpotato whitefly at bay. "When we get a pest and we spray insecticide trying to kill that major pest," said Toscano, "we may create another pest by releasing it from whatever biotic factors kept it under control. This happens."

In other parts of America, the quest to annihilate the boll weevil demonstrates emphatically the power of pesticides to elevate a pest from minor to major status. The tale of the boll weevil is told in *Ecologically Based Pest Management,* a book produced by the National Research Council. In 1950, after cotton growers had suffered the depredations of the boll weevil for half a century, DDT and other newly invented chlorinated-hydrocarbon compounds finally knocked down the champ of exotic insect pests. But in 1954, the boll weevil began showing signs of resistance to DDT, so growers switched to organophosphates. The organophosphates killed boll weevils and a broad range of other organisms, including the natural enemies of cotton bollworm and tobacco budworm, two lesser exotic pests, which up until then were minor pests. Soon bollworm and budworm populations erupted. Growers turned back to DDT in an effort to suppress bollworm, but it soon became resistant to DDT and other chlorinated hydrocarbons. So growers switched back to organophosphates, but again the pests soon developed resistance. By the late 1960s growers commonly called in 15 to 18 pesticide strikes a year and still weren't able to adequately protect their crops. A researcher estimated that one-third of all the insecticides used for agriculture ended up on cotton fields. Thanks to pesticide use, those once-trivial species, the cotton bollworm and tobacco budworm, had come to rival the mighty boll weevil as cotton pests.

The term resistance refers to the ability of pest species to select for genes that resist the toxic effects of a pesticide, thus creating populations that are immune. (The same mechanism accounts for the growing number of human disease organisms that defy antibiotics.) Resistance has become the bane of chemical pesticide manufacturers and users. Since the mid-1950s—about a decade after Americans began their love affair with synthetic chemical pesticides—the cases of resistance began mounting rapidly. By 1990, some 500 arthropods (insects, spiders, and kin) and a hundred plant pathogens had become resistant to at least one chemical. Despite efforts to avoid resistance, that growth curve continues to angle up the graph. As manufacturers exhaust the easiest options, the replacement by pest-resistant chemicals becomes increasingly expensive and difficult. One researcher calculated that resistance adds about $400 million a year to the expense of pest management.

After several years of frantic research, the chemical industry delivered several products that killed silverleaf whiteflies without known undue side effects. But the specter of resistance has taken some of the shine off these recent advances. The most effective products, for the majority of crops, were insecticides using imidacloprid, notably a pesticide called Admire. The EPA granted an emergency clearance, and by 1995 growers were using Admire all over the Imperial Valley and throughout whitefly territory. Imidacloprid has helped get Imperial Valley agriculture back on track, but how long will it be before resistance derails it? "I think it's only a matter of time," said Natwick, "before we lose Admire and could have a flare-up again."

Resistance, cost, the killing of beneficials—pesticides carry quite a burden. There are also the long-term damaging effects to the environment and to human health. Many farmers are beginning to realize that synthetic chemicals are not a silver bullet for pests, but it's hard for them to relinquish this appealing notion, which the chemical industry aggressively sold to their parents and grandparents. For the foreseeable future, pesticides will remain an important tool, and for at least a few years, they will remain the weapon of choice. It appears that in the long run, learning to live with pests means learning to live by the tenets of a more progressive approach—integrated pest management (IPM).

IPM practitioners control pests by adapting agriculture to natural processes and by using biological tools; chemical pesticides get relegated to the status of last resort. IPM appeared in the 1950s as "integrated control," though its common sense essence dates back thousands of years. Unfortunately this century's false promise of quick remedies—in this case chemicals—elbowed IPM to the periphery as a method of pest control. Only recently, as the shortcomings of chemicals have become painfully obvious, have age-old approaches such as those incorporated into IPM moved into the mainstream. But it's got a long way to go. Many people pay lip service to IPM, but most growers who profess to use IPM merely adhere to economic thresholds and try to use chemicals in ways that don't do as much harm to beneficials. Important steps but hardly a full embrace of IPM.

Prompted by desperation during the silverleaf whitefly crisis, many growers in the Imperial Valley did venture a little farther down the IPM path. They lobbied for whitefly-resistant cultivars and stepped up their pursuit of biological control. Larry Gilbert is among the growers who believe biocontrol is critical. "Without it I think our days are numbered. You can't live with something like [whitefly] unless it has more natural enemies. Whitefly are just so prolific. And they have the potential to not just harm but to destroy a crop. You can't win. We don't have the ability with just raw power to obliterate insects." Gilbert came up for air, then added, "We have to have enough beneficials to keep them moderately in check. Then if once in a while they get out of hand, we can control that [with pesticides]."

Most significant, some members of the agricultural community began to look at their area as an ecosystem, which is the prime directive of IPM. When a pest invades, one must understand the ways in which the agroecosystem encouraged that invasion and how the system can be used against the pest. Agricultural Commissioner Steve Birdsall said, "The buzzword now is 'integrated pest management,' but it should be 'integrated crop management' because it is not just the pest you're trying to manage but the total environment." Scientists point out that the artificial character of an irrigated desert chock-a-block with exotic crops is the root of the problem. Jon Allen, an ecologist at the University of Florida, has been studying the Imperial Valley and the relationship between its farming processes and the whitefly outbreak. "It's an ecological disaster that we somehow caused," said Allen. "Somehow we're turning Colorado River water into whiteflies."

By examining the valley as an agroecosystem, researchers and growers realized that whiteflies reached epidemic levels in part because they built up their numbers virtually uninterrupted from spring through fall. The whitefly population would hardly break stride following the harvest of the crop on which the whiteflies had been feeding; often they survived in the post-harvest debris until they could move on to the next crop. The Whitefly Committee, which growers formed in 1991, made the clean-up of residue one of its first initiatives. "As soon as you're through harvesting a crop that is susceptible [to whiteflies], get the damn thing disked up," said John Pierre Menvielle, the second-generation valley farmer who is a charter member of the committee. "Don't leave it so the whitefly can repopulate." Birdsall noted that his office had the authority to order growers to eliminate crop residues, but he rarely had to use it. Most farmers readily complied; those who were slower to come around got phone calls from their fellow growers. "What we used most successfully was peer pressure," said Birdsall.

"What your neighbor does affects you," said Gilbert, "especially with a pest that may travel a few miles. So if melon growers and broccoli growers don't destroy their crop residue in a timely manner after harvest and the whiteflies develop into huge numbers, it's going to affect the people downwind and in the general area. So I'm inclined to go for an area-wide solution." Gilbert knows that addressing problems as a region is a key to IPM, but he warns that getting all the farmers in the valley to follow certain requirements is problematic. "Most people here feel that we already have enough rules and requirements and restrictions and regulations—I call them the four R's," said Gilbert, smiling wryly. "Farmers like to think they're independent. We're not, but we like to think that. We don't particularly like additional regulations, so you want to be real sure that it's needed and that it'll actually do some good."

Some farmers never bought into IPM, though, and others promptly jilted it as soon as Admire showed up and gave a wink. "I think there's a feeling out there," said Birdsall, "that the whitefly isn't a problem anymore." Many farmers are grumbling about paying for the Whitefly Committee and the research it supports. "We have some resistance now because the whitefly is not as evident," said Menvielle. "Some growers think we're wasting money." But he strongly disagrees: "If we're not working on research, we're eventually going to get killed. I think the people who are going to be competitive and survive and be successful are those who are progressive and look down the road to these long-term solutions."

For that matter, the short-term battle is far from over. Larry Gilbert is one grower who doesn't believe for a minute that the whitefly is history: "Any crop that's susceptible, that isn't protected well, is going to be in trouble." Even the guys down at the Central Barber Shop in El Centro see ongoing problems. "You don't even dare plant tomatoes in your backyard," said Mel Valenzuela, the older barber. "They love tomatoes. They suck the juice out of them." Al Payne, Mel's customer, added, "It's still bad out there. I know lots of places, when I go walking in the morning, there are clouds of whiteflies along the ditch banks." Entomologist Steve Castle thinks valley residents who now dismiss the whitefly have lost their perspective. "We're complacent about it because we have the comparison to the early years, but compared to anywhere else . . . we're still thick in whiteflies. This valley is still whitefly city, and it always will be because of agriculture."

Castle probably is right. The valley likely will remain whitefly city—and a stronghold of numerous other alien pests—unless the revolution in farming practices comes to pass. Until then, the bugs will dictate much of life in the Imperial Valley. "Some of these pests," said Menvielle, "we're just going to have to learn to live with them, farm around them, because they're not going to go away. And who knows what's going to be coming down the road?"

Menvielle notes that the Imperial Valley needs not only area-wide cooperation but international cooperation. Bumping down a farm road on his property, he and I passed within just a few hundred yards of Mexico, the border fence easily visible. "The whitefly doesn't need a green card," said Menvielle. "No fence is going to stop him from coming across." Back in 1994, when farmers largely had controlled the whitefly in cotton in the Imperial Valley, they still had to battle whiteflies because they poured over the border from the Mexico. No doubt American-born whiteflies likewise drifted across the border into Mexico in other cases. Fortunately Mexicans have participated in the campaign against the whitefly from the start and the transnational whitefly flow has abated.

Unfortunately another effort to keep a winged pest from crossing the border didn't work out so well.

I watched in horror as 71-year-old Frank Garcia stumbled around his yard with a hundreds of killer bees on his face. My horror stemmed partly from Garcia's grisly fate and partly from the unabashed sensationalism of this re-created tragedy, which a television network broadcast in 1997 as part of a program demurely titled, "The World's Deadliest Swarms." Killer bees are probably the most famous exotics in the United States. More properly called Africanized honeybees, they're a hybrid of African honeybees and European honeybees. In an attempt to boost honey production, researchers in Brazil imported African honeybees in the 1950s, hoping to breed bees that featured the most desirable traits of both the African and European types. Unfortunately in the late 1950s, some African honeybees broke out of the research facility. They began mating with European honeybees in an unscientific fashion, and the eventual result, Africanized honeybees, began dispersing throughout the warmer climes of the western hemisphere.

As Africanized honeybees moved north through Mexico, their pending arrival in the U.S. became grist for the media mill (hence the seventeen *Saturday Night Live* killer-bee sketches, the first of which aired in 1978). The real bees finally debuted in October 1990, crossing the border into Hidalgo, Texas. They've since spread to about a third of Texas and to the southern parts of New Mexico, Arizona, and California. They seem to be running out of steam, however, and apparently won't inhabit as large an area as first predicted. Ironically, they're probably being thwarted at least in part by parasitic tracheal and varroa mites—both fellow exotic species—which in recent years killed nearly every European honeybee in the U.S. that didn't live under the protective care of beekeepers.

So, really, do killer bees kill? Yes, but very rarely, according to the researchers at Texas A&M University, which serves as a clearinghouse for information on Africanized honeybees. The university's spokesperson on bees, Kathleen Davis, reports that since the bees entered Texas in 1990, they have caused two deaths in the state. One victim was a 96-year-old man who died from a heart attack two days after being stung. The other person died after being stung only 40 times, which indicates he was allergic. Researchers have calculated that it takes about ten stings per pound of body weight in order to inject enough venom to kill a nonallergic person—that would be about

1,500 stings to bring down the average-size adult male or female.

Though their stings rarely kill, Africanized honeybees shouldn't be taken lightly. They do attack people more fiercely than do regular honeybees, though Davis cautions that the European honeybees deserve respect, too. Texas A&M provides tips for residents and travelers in killer-bee country. Some examples: the vibrations and exhaust from lawn mowers, chain saws, and other machinery can provoke a bee assault. If Africanized bees assault you, don't swat at them; when they're crushed they release pheromones that signal others from their hive to attack. European honeybees on average will chase someone about 50 yards in order to run the interloper out of their territory; Africanized bees will chase someone from 100 to 150 yards. Duck into a building, a car, or even thick bushes, but don't dive into a pool or lake; the bees usually will be waiting for you when you come up for air. If the bees catch you out in the open, run. The bees fly 12 to 15 miles an hour, which is a little more than half the speed of an Olympic sprinter in full stride. That's pretty fast, but think how motivated you'd be.

Fire ants are easy to outrun, but due to sheer numbers they cause far more trouble than killer bees do. Two species of exotic fire ants make trouble in the U.S.: the red and the black imported fire ant. The red species causes the vast majority of the problems and is the culprit to which I refer when I use the generic term "fire ant."

Both species entered the country via dry ship ballast at the port of Mobile, Alabama; the black ant in 1918 and the red around 1930. Fire ants expanded quickly through the warm, moist Southeast; the leading edge of the infestation grows about 20 to 30 miles a year. Today the ants dominate about 260 million acres, encompassing most of nine southeastern states and Puerto Rico. Fire ants also have shown up in California, Arizona, New Mexico, Kansas, and Maryland.

People despise and fear fire ants because of their bites, which are in most regards worse than the bites of native ants. Fire-ant bites burn painfully, they often cause infections, they produce ugly white pustules, and they can leave permanent scars. Fire ants drive anglers from the banks of rivers and lakes; they ruin backyard barbecues; they can render a campground uninhabitable; and they elevate the legendary conflict between ants and picnickers to new heights. Because fire ants swarm quickly over their victims and each ant can and does bite repeatedly, they do kill people, though even more infrequently than do Africanized honeybees.

In 1997, the Texas legislature passed a fire-ant law that provided funding for an extensive control program. "It's a great day in Texas, folks," proclaimed State Representative Tom Ramsay, the driving force behind the legislation. "Mothers in towns all over Texas are once again going to be able to let their kids go out in the backyard to play again and not be afraid of what the fire ant is going to do to that child."

I recently spent some time in Florida with Greg Jubinsky, who works for the state Department of Environmental Protection. Though Jubinsky's passion is fighting invasive exotic plants, he harbors a special malice for fire ants—they almost killed his son Drew when he was two. While playing in the backyard, Drew sat down on a fire-ant hill. The ants swarmed over him almost instantly and bit him all over before Jubinsky's wife, Karen, ran over and snatched him off the hill. Drew lapsed into unconsciousness and his tongue swelled, threatening to suffocate him. Karen, a nurse, administered first aid and rushed him to a hospital, where he recovered. Such incidents aren't uncommon enough in fire-ant country.

For as long as fire ants have been around, people have been trying all sorts of things to kill them. Pesticides remain a staple, despite the disaster that followed their misuse in the fire-ant eradication attempt 40 years ago. In urban areas in Texas alone, homeowners spend about $90 million a year fighting fire ants, and most of that is spent on chemicals. People try many home remedies, too, pouring all sorts of stuff down fire-ant holes, including motor oil, gasoline, bleach, and battery acid. Some individuals tout instant grits as the answer to fire ants. The theory goes that ants will eat the grits and then the grits, upon contact with liquids inside the ants, will swell and rupture the ants' stomachs. It doesn't work.

Whether someone uses instant grits or sophisticated chemical pesticides, fire ants resist control. Their primary defense is fecundity. One colony can produce thousands of reproductive females, each of which, after mating with a male, can start a new colony wherever it may land. These queens also enjoy a protection that one usually associates with medieval kings: they have royal tasters. Pesticide designers figured that poison baits would work great with fire ants, because the workers would drag the bait down into the hole to the queen, whose throne room often is too far underground to reach by other means of application. But it turns out that queens only eat food that is first eaten by a worker; and if the worker dies, the queen doesn't touch it.

In addition to biting and stinging humans, fire ants and killer bees also do much damage to agriculture and the environment. The U.S. Department of Agriculture estimates that Africanized honeybees cost American beekeepers about $29 to $58 million a year, mostly by competing with European honeybees for nectar and by interbreeding with the European bees and producing offspring that are more difficult for beekeepers to manage. Fire ants do far more damage. They wreck farm equipment, irrigation systems, buildings, and electrical appliances, such as air-conditioners. They feed on seeds, fruits, and young plants. They erect big, hard mounds in farmers' fields that make mechanical harvesting difficult, and they bite workers who harvest crops by hand. They injure and sometimes kill animals, especially calves; in Texas alone fire ants cost the cattle industry about $67 million a year. Fire ants also degrade the environment. When they sweep into an area, they obliterate most native ants, plus a good many spiders, worms, and other invertebrates. Scientists don't yet thoroughly understand the ramifications of losing all these little creatures, but pessimism is justified.

Fire ants are not the only exotic pests that harm natural areas. Though we usually think of pests as a menace to agriculture, many of them inflict tremendous damage on the flora and fauna of our wildlands. The alien pests that harm some of our native trees illustrate this problem all too well.

Since 1930, Dutch elm disease has travelled from coast to coast, killing elms by the hundreds of millions. Most of us know of Dutch elm disease as a fungus that leveled the trees lining the streets of our cities and suburbs, but most of these victims were European elms. From the Atlantic to the Great Lakes, Dutch elm disease killed nearly all of the approximately one billion indigenous American elms that once occupied the northern forests.

The European gypsy moth rivals Dutch elm disease as a destroyer of forests. Gypsy moths erupt periodically but unpredictably, rampaging across millions of acres in summertime blitzkriegs, defoliating pines, beeches, blueberry bushes, hemlocks, apple trees, birches, and, especially, oaks—about 400 species of trees and shrubs altogether. Gypsy moths first appeared in the U.S. in Massachusetts, shortly after the Civil War. They've spread throughout New England and contiguously westward as far as the Great Lakes and southward as far as Virginia. They've also jumped to Colorado, Utah, and all three West Coast states.

Perhaps the most destructive exotic tree pest in our history is chestnut blight, the fungus *Endothia parasitica*. It's not a household word anymore, like Dutch elm disease and gypsy moth, because it ran out of trees to kill a couple of generations ago: chestnut blight killed virtually every American chestnut tree, a species that was abundant on well over 200 million acres of eastern forest.

In their 1967 book about plant disease, *Famine on the Wind*, G. L. Carefoot and E. R. Sprott chronicle chestnut blight's invasion of the U.S. One of the earliest reports of the blight occurred in 1904 in the New York Zoological Park. Herman Merkel, the zoo's chief forester, noted with concern that the chestnut trees in the park were wilting and dying. His attempts to save them failed and eventually he had to cut down all the trees. But no one felt alarmed because the mysterious disease seemed to be confined to the zoo grounds. But not for long. By the next year, reports of the blight streamed in from all over the state of New York. Steadily the fungus spread north, south, and west, killing every chestnut in its path. The U.S. Department of Agriculture set up a special laboratory to study chestnut blight and states set up well-funded anti-blight programs. Nothing helped. By 1915, all the government efforts shut down and people watched helplessly as the blight rolled west as far as Michigan and south as far as Louisiana—the main range of the American chestnut. The savory nuts referred to in the opening line of the classic Christmas song "Chestnuts Roasting on an Open Fire" had vanished into memory by the time Mel Tormé wrote those lyrics in 1946.

The fungus kills a chestnut tree by starving it. Just under the outer bark lie the cortex cells, where food is stored, and the phloem, which consists of millions of slender tubes that transport food up and down the tree. The fungus has to find a wound in order to enter, but almost any tiny nick will suffice, from the hole of a boring insect to the scratches left by a squirrel. The fungus slips the thread masses that make up its body into the cortex tissue and phloem, and then it feeds on the tree, depleting the victuals stored in the cortex cells and clogging up the tubes of the phloem. Eventually the fungus works its way all around the inside rim of the twig or branch and chokes it off, killing the part of the tree beyond the tourniquet. When the tentacles encircle the insides of the trunk, another withered American chestnut crumples to the ground.

As the fungus feeds and prospers, it pushes hundreds of pinhead-size fruiting bodies out through the bark. During wet weather, special fungal threads called spore horns emerge through the fruiting bodies. The spore horns disintegrate into millions of sticky spores that go hunting for new chestnuts by washing away in the rain or by clinging to the feet of insects or birds. The fungus also spreads by forming millions of little sacs, each containing eight spores. When the sacs reach the surface of the bark, they burst and launch the spores into the air, where the wind carries them to chestnut trees miles away.

Chestnut trees are more than the stuff of Christmas carols. Durable and resistant to dry rot and weather, chestnut wood was prized by builders and woodworkers. Carefoot and Sprott report that the blight wiped out about 30 billion board feet of chestnut lumber in Appalachian forests alone, where about 25 percent of the trees were chestnut. The fungus also shut down a number of large tannin-extracting businesses in Appalachia: the heartwood of American chestnut trees contains a sizeable store of tannin, a chemical used in making leather. Though the American chestnut didn't survive long enough for its ecological role to be well understood, one can assume that the loss of all those trees and all those tasty nuts had repercussions. Carefoot and Sprott assert that populations of squirrel, deer, and wild turkey declined dramatically, and the OTA report notes that at least five species of native insects are thought to have disappeared along with the chestnut.

The ecological consequences of a tree-killing pest are better known in the case of the balsam wooly adelgid. This European aphid appeared in the U.S. in the early 1970s and so far has spread through Virginia, North Carolina, and Tennessee. It is partial to firs, which it kills over the course of several years by sucking out their sap. The balsam wooly adelgid may work slowly but it is thorough; it has destroyed nearly every adult fir tree in Great Smoky Mountains National Park, which had been the haven for three quarters of the spruce-fir forest left in the southern United States.

A Nature Conservancy report chronicles the cascading impacts set in motion by the annihilation of the firs in the park and elsewhere. As the firs turn into skeletons and topple, the forest loses its dominant canopy tree and becomes warmer and drier. The rising temperatures threaten a number of ice-age relics—northern plant and animal species that have survived for thousands of years in this southern pocket of cool, high-elevation forest. Other species that appear to rely exclusively on firs for food, such as the native moth *Semiothisa fraserata*, likely will falter. Eight species of rare mosses and liverworts that attached to the bark of fir trees may not be able to adapt to other homes. Most native insect species can meet their needs by turning to the red spruce, but there simply aren't as many spruce as there were firs. A few insect species could move to lower elevations and use hemlocks, but hemlocks are falling to another exotic pest, a cousin of the balsam wooly adelgid called the hemlock wooly adelgid. This loss of habitat has harmed local populations of hermit thrush and northern saw-whet owl and has been the main factor in driving the Carolina flying squirrel, spruce-fir moss spider, and rock gnome lichen onto the federal endangered-species list. Currently no one knows how to stop the balsam wooly adelgid.

In the Nature Conservancy report, Dale Schweitzer, a Conservancy zoologist, says, "The only apparent hope for recovery of these forests seems to be a combination of one or more successful biocontrols—and a few centuries of healing time."

CHAPTER 7

ORCHIDS, EGRETS, AND WILD LIMES: THE ENDANGERED EVERGLADES

There are no other Everglades in the world.

They are, they have always been, one of the unique regions of the earth, remote, never wholly known. Nothing anywhere else is like them: their vast glittering openness, wider than the enormous visible round of the horizon, the racing free saltness and sweetness of their massive winds, under the dazzling blue heights of space. They are unique also in the simplicity, the diversity, the related harmony of the forms of life they enclose. The miracle of the light pours over the green and brown expanse of saw grass and of water, shining and slow moving below, the grass and water that is the meaning and the central fact of the Everglades of Florida. It is a river of grass.

Marjory Stoneman Douglas got it right in those opening words of her 1947 classic, *The Everglades: River of Grass.* The central fact of the Everglades is indeed saw grass and water. By extension, one can say that aquatic plants and water are the central fact for all of South Florida. But heedless development has disrupted these essential elements, and has been doing so since before Douglas penned her seminal book. She chronicled past destruction and warned of future harm, and that harm has come to pass. In the five decades since the publication of *The Everglades: River of Grass,* much of South Florida's signature landscape of plants and water has been paved, plowed, plumbed, and polluted.

For the last half century, Douglas, now 108, has continued fighting for the welfare of her beloved Everglades. Many others have joined the long struggle to protect and restore South Florida's watery habitats, yet only in the last few years have their efforts won substantial victories. Unfortunately an explosion of water-loving exotic plants has darkened this brightening future.

The threat of invasive plants wasn't on the radar screen when Douglas wrote her book. She makes only offhand references to the presence of Australian pines, hyacinth, bamboo, and "careless weeds" in towns and along roads, but with no sense of foreboding. Little did she, or anyone else, suspect that a confluence of factors eventually would turn South Florida into the aquatic weed capital of the nation.

For starters, South Florida teems with the aqua that aquatic plants need. (I'm using "aquatic" broadly to encompass all watery habitat, from lakes to seasonal wetlands.) Flying over South Florida on a sunny day, I saw the sparkle of water in all directions, despite the extensive draining that has occurred in the twentieth century. Some of that sparkle came from a few of the state's 7,800 lakes and 1,700 rivers. Less visible from the air but even more prevalent are millions of acres of wet prairies, swamps, and marshes, the kinds of watery places that once defined Florida.

In ecological terms, South Florida is something of an island, bounded on three sides by water and on the north by the frost line, which runs roughly from Daytona Beach through Orlando to Tampa. As an isolated almost-island cut off from much of the continental flow of plants and animals, South Florida harbors a relatively low diversity of native species. Because it has fewer species that can compete successfully with invasive non-natives, exotics thrive in the region.

The tropical climate of South Florida also contributes to its status as the nation's aquatic-weed capital. The world's tropics harbor by far the most diverse assemblage of flora and fauna on Earth, which means there are millions of species prowling the planet that might prosper in the warm, wet climes of South Florida should they get there.

And many are getting there. Every year government agencies, botanical gardens, home gardeners, and the agricultural, nursery, and aquarium industries bring into Florida millions of plants from overseas. Millions more destined for places all over the U.S. pass through the state, increasing the odds that the seeds or spores of invasive plants will escape and become established in Florida. South Florida's popularity among travelers and immigrants adds even greater possibilities for exotic-plant introductions. Some Asian immigrants, for example, grow water spinach, a highly invasive aquatic morning glory that is widely prized as a vegetable in parts of Asia. Because it's illegal to plant or possess this federally prohibited weed, water spinach is marketed covertly; it's even sold out of back rooms that one cannot enter without providing a secret password.

Given the circumstances, it's not surprising that more than 900 species of non-indigenous plants are growing clandestinely outside cultivation in Florida. They constitute about 27 percent of the state's flora, and that percentage continues to grow. The slow workings of evolution create a new native species only about once every eon, yet new exotics keep showing up. The majority of the newly established species have taken root in South Florida, and most of them have shown up in aquatic communities.

Within hours of arriving in Florida I met hydrilla, arguably public enemy number one among exotic plants that invade open water. This import originated in Sri Lanka and appeared in Florida waterways in the 1950s. It has spread to nearly half of the state's public waters. It is now by far the most abundant open-water plant in Florida. Between 1992 and 1994, the amount of public water infested with *Hydrilla verticillata* doubled to about 100,000 acres—and no one knows what's happening on the water bodies the state doesn't survey, which includes almost all of Florida's small and medium-size lakes.

According to the Army Corps of Engineers, hydrilla also plagues the lakes, ponds, and canals of 20 other states. Resource managers from California to Connecticut spend millions every year in an attempt to control hydrilla; given that it makes itself at home in many different environments, the plant eventually could expand throughout most of the U.S.

To give me a close look at hydrilla, Ed Harris rummaged through the air boat's storage bin and pulled out the frotus. Harris is a biologist who works out of the Orlando office of the Bureau of Aquatic Plant Management, an agency within the state's Department of Environmental Protection. Also on board was Greg Jubinsky, a bureau manager and a leader in the war on invasive weeds in Florida. Our air boat was bobbing on Lake Kissimmee, a 35,000-acre body of water that lies about 30 miles south of Orlando, near the top of South Florida. Here I began my journey through the realm of invasive aquatic plants, from the frost line down to Everglades National Park, which occupies the southernmost tip of the Florida mainland.

The frotus turned out to be a sort of grappling hook; it looked like a double-sided rake head tied to a rope. Grinning through his dark full beard, Jubinsky, who has lived in Florida for virtually his entire life, explained that the now-official name for this implement derives from the way in which a down-home Floridian would drawl "throw this."

Harris threw the frotus into the water and hauled in a thick, dripping, green-brown tangle of hydrilla. He held it up like a trophy marlin. And quite a catch it was; its tendrils spilled from a height of six feet down to the deck. Such masses intertwine to form a dense mat atop the water—about 70 percent of a hydrilla plant typically bunches up on or near the surface. "You'd swear you could walk across it," said Jubinsky. In fact, animals as large as ducks do walk on it. Because hydrilla is not a free-floating plant and must root in the soil at the bottom of the lake, one would expect to find it only in the shallows close to shore. Most lakes are too deep for rooting plants; they can't grow hundreds of feet below the surface. However, most of South Florida's lakes are shallow throughout. Though Lake Kissimmee is a very large body of water, it only reaches depths of 12 or 14 feet. Hydrilla can grow in water as deep as 20 feet, and in exceptional circumstances to a depth of 50 feet. Entire lakes have been smothered by hydrilla like a lid on a pot. And when colonizing new water bodies, hydrilla enjoys several advantages over most native aquatic plants. It starts growing earlier in the year and earlier in the day than its indigenous competitors. It can also root in deeper water than most of South Florida's native plants due to its rare ability to grow at light intensities as low as one percent of full sunlight. To top it off, hydrilla can grow at the remarkable rate of several inches a day.

Sprawling jungles of hydrilla interfere with flood control, boating, fishing, swimming, waterskiing, and other human uses, but biologists worry most about hydrilla's effect on the aquatic environment. "It'll grow up from the bottom," said Harris, "and spread along the top of the water, almost like a big umbrella, and it'll shade out the native vegetation underneath. It reduces the overall diversity of species out here. . . . Eelgrass and pond weed have suffered significantly." As for other native organisms, Harris isn't sure how they're faring. He says funding for research has been slim and it's hard to say how much the ecosystem is slipping, or what the impacts might be 50 years down the road. "These are dynamic systems," said Harris. "Small changes can have big impacts." And he emphasized that hydrilla has caused much more than a *small* change.

Five Florida-based scientists recently reviewed the literature on hydrilla and found ample reason for concern. Writing in *Strangers in Paradise,* a 1997 book about invasive exotics in Florida, they note that in addition to reducing light penetration, hydrilla alters the dynamics of oxygen transfer and retards water circulation. They add that beneath the canopy of hydrilla, "extremely low levels of dissolved oxygen and significant changes in turbidity, color, cholorophyll, pH, alkalinity, specific conductivity, and phosphorus are common." Finally they point to research linking hydrilla to shifts in the density and diversity of zooplankton, epiphytes, and bottom-dwelling invertebrates.

Convincing state legislators to fund research on zooplankton—minute animals they've never seen—presents difficulties, but hydrilla attracts some public dollars because it can seriously reduce recreation revenues and, by clogging channels, impede flood control. "Economic concerns, that's what gets us money," said Harris. "We tell legislators, 'You know, if this lake fills up with hydrilla, people aren't going to come from all around the world to go bass fishing.' One of the hottest things in Japan right now is to come to Florida and spend a week bass fishing, spend a week playing golf, and do the theme parks. Japanese couples come here on their honeymoons. They'll spend five solid days bass fishing on this lake. They're paying a fishing guide a thousand dollars a day and staying in hotels. You can show there's a direct economic benefit in managing these lakes." A study of two connected lakes in central Florida found that the lakes add about $10 million a year to the economy, almost all of which would be lost if the infestation of hydrilla expands throughout the lakes.

But the hydrilla-economics equation hasn't always worked out as one might expect. Clearly, fisheries managers don't want to see lakes smothered by hydrilla. Nor do they want hydrilla to reach densities of 30 to 50 percent of the area, which stunts the growth of some prized game fish. But many fisheries managers want to have some hydrilla around because they think that it enhances game-fish habitat and that fish congregate around hydrilla, making them easier for anglers to locate. Duck hunters also like to have some hydrilla around; in addition to walking on it, ducks eat hydrilla.

Greg Jubinsky, however, doesn't approve of granting even partial legitimacy to hydrilla just for the sake of recreation. "Hydrilla is an artificial substrate [the foundation on which an organism attaches and grows] for invertebrates, fishes, and the like. Because it's an artificial substrate you're going to have increased levels of nutrient interactions in the lake. You're going to get . . . an artificially exaggerated sport fish population growing in relation to an artificially inflated plant community . . . but I don't think these lakes are here . . . to be bass factories." Harris chimed in: "Hydrilla is good for bass only because having substrate below the surface is good for bass. If it was any other plant besides hydrilla, it would be just as good." Harris points out that bass, ducks, and other game species did just fine in these lakes long before hydrilla showed up. "Our goal as an agency is to try to get these things back to the way they were before Europeans showed up. If we can do that, the lakes will provide all these things on their own."

After grappling with the hydrilla, Harris stowed the frotus and we climbed up into the elevated seats of the air boat. As Harris switched on the engine that drives the wading-pool-sized fan mounted on the rear, we donned our heavy-duty ear protectors (the roar of the fan is so loud that anglers 100 yards away covered their ears with their hands) and we sped out, skimming swiftly over open water. The wind created by our boat's speed was a welcome respite from the muggy, 90-degree summer heat.

After several minutes, we entered a vegetated area near the shore and Harris eased off the throttle. A smallish alligator slid silently out of our path. Some great egrets rose from the shallows, their whiteness vivid against the blue sky. Another gator, also of modest size, lay motionless atop the water, a log with wary eyes. Tricolored herons, snowy egrets, and great blue herons minced about in ankle-deep water, spearlike bills poised to strike should a fish come within range.

Harris stopped the boat close to the shore and cut the engine. Now that conversation was again possible, he explained that the open water we'd just crossed had been choked with hydrilla just a few years before. It had covered more than a third of the lake. Controlling it was not easy or cheap, he pointed out. And just getting a handle on the extent of hydrilla infestations still remains a problem, given the fact that hydrilla can double in area in as little as two weeks when left to its own devices. In August 1987, surveyors counted about 4,000 acres of hydrilla in Lake Istokpoga, a 28,000-acre lake located about 30 miles south of Lake Kissimmee. By December 1988, the infestation had ballooned to nearly 20,000 acres.

Aside from measuring the spread of hydrilla, what can resource managers do? Early on, managers tried to remove hydrilla with mechanical devices, such as harvesters and draglines, which they hauled across the bottoms of lakes and waterways. This proved ineffective, destructive, and expensive. Hydrilla grows back so rapidly that these mechanical means have to be employed two, three, even four times a year, costing from $2,000 to $3,000 an acre for each treatment. Native plants and animals die in the process, and the death of native competitors sometimes hastens the spread of hydrilla. So does chopping it up. Jubinsky dropped a one-inch piece of hydrilla into my hand and told me that even such a tiny fragment could start a new hydrilla infestation by hitching a ride to an uninfected body of water. "The main way hydrilla spreads is by boats and boat trailers," said Jubinsky. Showing respect for a worthy adversary, Jubinsky added, "You have to admire hydrilla as a plant."

Some water managers turned the power of plumbing against hydrilla. Using South Florida's vast network of canals, pipes, and pumps, they temporarily drained bodies of water, a process called a drawdown. Drawdowns kill the standing crop of hydrilla, sure enough, but they also kill most other aquatic plants and animals. In addition, boating enthusiasts, anglers, bait-shop owners, and water-front residents tend to frown on the periodic draining of their lakes. Besides, hydrilla has a trick up its sleeve. An average hydrilla plant produces several thousand pea-size reproductive structures called tubers and turions that can remain viable indefinitely in the moist soil left behind. When the water returns, so does hydrilla, resprouting from its underground hideouts.

Hydrilla fighters rely mainly on herbicides, though using chemicals in aquatic systems is touchy business. It took a long time to find herbicides that the bureau considered effective, reasonably priced, and not too dangerous; government agencies evaluated some 800 compounds before settling on only a handful. Even some of those chosen had drawbacks, most notably their effects on native organisms. As Jubinsky said: "You'd put [the chemicals] into the water at the front of the boat and fish would already be floating when the back of the boat went by." But Jubinsky and Harris think that in recent years the chemical industry has developed some safe and effective hydrilla killers. These bureau biologists especially like fluridone, which they've been applying to Lake Kissimmee. Harris says that fluridone deserves much of the credit for safely clearing hydrilla from the large area of open water we'd skimmed across in the air boat—that is, fluridone *and* the $2.5 million it took to buy and apply it.

Funding has become the key in the war against hydrilla, and is often the key in controlling exotic invaders. When resource managers know how to control exotics, they then need the dollars to implement control measures. Even allowing for the fact that government agencies lust for additional funding the way a bear desires honey, a bigger budget for the control of exotics would definitely make sense—and the sooner the better. The history of hydrilla management in Florida provides well-documented support for the necessity of adequate funding.

The Bureau of Aquatic Plant Management has made the early eradication of new and small populations its top priority in controlling hydrilla. Since 1990, bureau managers have erased or reduced to insignificance several dozen pioneer colonies for about $5,000 each. In 1994, the bureau allocated about $450,000—a substantial sum—to eliminate or sharply reduce about 160 existing infestations of less than a hundred acres each. When it came to take the final, crucial step in the well-planned management program—tackling the additional 40-plus bodies of water saddled with large hydrilla populations—the legislature tightened its purse strings. The bureau's coup de grâce could have successfully reduced hydrilla to just another background weed.

The bureau calculated in 1995 what it would take to deal with the 40-plus major hydrilla infestations: $13 million a year for two to three years. The bureau had in mind the successful program employed in Lake Lochloosa as a model. For a decade virtually every one of this lake's 5,700 acres had languished beneath a thick pelt of hydrilla. Then in 1993, the bureau made Lake Lochloosa a showcase for controlling large-scale hydrilla strongholds. Massive herbicide treatments that year and in 1994 reduced the standing crop of hydrilla to one acre. Those two years of all-out attack on a huge infestation cost about $1.25 million, but from now on the annual cost will run much lower, fluctuating between several hundred thousand dollars and zero depending on when the bureau has to deal with new growth. And there always will be new growth. Hydrilla sprouting from those chemical-proof tubers and turions reaches maturity in mere weeks, at which time the latest progeny produce tubers and turions themselves. Given current technology, once hydrilla becomes established, to some extent it will be around forever.

The Lake Lochloosa example of forward-looking management did not loosen the legislature's grip on its wallet. Instead of meeting the bureau's requests for around $13 million a year, the legislature provided from 1992 to 1996 about $4 to $7 million annually. Positive thinking and hopeful arithmetic might suggest that the bureau could have still achieved its goals with this funding, that it simply would have taken a few more years to get around to all the large infestations. But while managers were subtracting sizeable amounts of hydrilla from Lake Lochloosa, large amounts were being added to bodies of water that didn't get the full, necessary treatment.

Planning and prevention that can reduce long-term management costs seems like an approach any government would appreciate, but politics interfered with the bureau's efficient use of its already inadequate budget. Citizens of Florida, particularly sport-fishing enthusiasts and allied businesses, demand access to public waters. Every year the Bureau of Aquatic Plant Management must devote a significant portion of its budget to clearing just enough hydrilla from all public waters to ensure access for fishing and other forms of water recreation. These minimal treatments do no lasting good and must be repeated year after year. In the long run, the legislature's incomplete approach has cost the economy and environment far more than would an expensive but brief campaign that reduced every hydrilla population to manageable levels.

It's easy to see why managers struggling with hydrilla in the mid-1990s found their situation frustrating. But it was doubly so because years ago the legislature funded the successful Lake Lochloosa approach to deal with water-hyacinth and water-lettuce invasions. Despite the horrors of hydrilla, the official highest priority for the Bureau of Aquatic Plant Management is controlling these two floating plants. This is surprising when one learns that together water hyacinth and water lettuce cover only about 4 percent as much area as does hydrilla—and these plant populations are no longer expanding. But there's a good reason.

Water hyacinth, the worse of these two similar invaders, came to America in the 1880s, imported by residents who admired its spike of lavender flowers and thick, waxy leaves. Able to double its population in as little as 12 days, this South American native choked Florida's St. Johns River within a few years of its appearance, disrupting steamboat traffic and logging operations. In 1899, Congress unleashed the Army Corps of Engineers on water hyacinth, but decades of effort by the Corps and others only slowed the expansion of this aggressive weed. By the 1960s, it still clogged more than 125,000 acres of Florida's waters, causing impacts akin to those of hydrilla.

Recognizing that water hyacinth was taking over Florida's waterways and that agencies were throwing good money after bad in erratic attempts to control it, the state legislature in 1974 adopted a policy toward hyacinth and other alien aquatic plants called "maintenance control." This policy directed various agencies, under the leadership of the Florida Department of Environmental Protection, to reduce populations of water hyacinth to the "lowest feasible level" and then keep them there. (The extra effort to eradicate these populations is not considered worthwhile because hyacinth will just move back in from unmanaged lands nearby.) The state and federal government provided major funding for the initial big push and ever since have consistently provided enough money to hold water hyacinth populations to only a few acres each. If they relented for even a few years, water hyacinth would explode. The state applied the same approach to water lettuce and enjoyed similar success. So in light of this exemplary precedent and the legal mandate from the 1974 law, bureau managers expected the legislature to fund the same strategy for controlling hydrilla.

One explanation of the Florida legislature's approach to the hydrilla problem relates to the point that Jubinsky and Harris made about anglers and duck hunters and their advocates in state agencies: many like a certain amount of hydrilla because it is thought to pump up fish and duck populations. Writing in *Strangers in Paradise*, Jeffrey Schardt, an environmental administrator with the bureau, offers the opinion that the legislators underfunded the hydrilla program because they had received mixed messages. While the bureau and others warned them of the evils of hydrilla, many in the duck-hunting and sport-fishing communities extolled the virtues of having some hydrilla. The opinion of anglers in particular carries enormous weight, because behind it stands the Florida sport-fishing industry, which, according to a 1997 study by the American Sportfishing Association, contributes more than six billion dollars a year to the state's economy. Such commercial self-interest hampers anti-exotics efforts around the country.

Schardt reports that, beginning in 1997, the legislature began appropriating a lot more money for hydrilla control: $11.1 million in 1997 and $12 million for 1998. This sudden increase no doubt stems in part from the relentless efforts of the bureau to educate legislators and the public about the dangers of hydrilla. But hydrilla itself bears most of the responsibility for this change of heart. By the mid-1990s, this nasty weed had covered so many lakes so thoroughly that sport fishers and waterfront home owners—two powerful constituencies—quit touting the benefits of hydrilla and joined the bureau in calling for more money for control. Finally given ample support, the bureau already has cut the number of acres of public waters infested by hydrilla from nearly 100,000 to well under 50,000. By the year 2000, barring hydrilla booms related to extreme weather, the bureau expects to have reduced hydrilla to maintenance control levels.

This prediction assumes that the bureau will continue to receive adequate funding for hydrilla, which is not guaranteed. The bureau's extra funding has been drawn from other environmental programs, such as recycling and wetland mitigation. More threatening, the coalition of interests that won more funds for hydrilla control has begun to fray. First of all, the duck-hunting lobby has always fought hydrilla control and continues to do so. More important, some of the sport fishers and home owners who recently supported control have begun to regress. As the quantity of hydrilla in their favorite lakes has decreased, they once again have begun trumpeting its virtue. Schardt just hopes that the bureau can get a handle on hydrilla before support wanes too much.

Perhaps some of the confusion over hydrilla would clear up if more people visited that part of Lake Kissimmee where Harris, Jubinsky, and I saw the gators, egrets, and herons. The area is part of a major restoration project. In addition to killing hydrilla, workers removed untold tons of muck that had accumulated on the bottom, unable to be flushed out because South Florida's tightly controlled water system stifles natural water circulation.

With most of the hydrilla gone, a healthy native community is returning. Harris pointed to the knotgrass poking from the water about a hundred yards offshore. This native aquatic grass doesn't get as thick as the alien species, allowing water and aquatic animals to circulate freely. Closer to shore, pickerel weed and arrowhead thrust above the surface. Harris said that waterfowl love arrowhead, commonly known as "duck potato," and that pickerel weed produces a showy purple flower. He added that both of these native plants serve as a substrate for snails and other invertebrates, which feed the prized largemouth bass as well as shiners, black crappies, bluegills, and other native fish. The fish in turn feed the gators and the birds that stalk the shallows around us.

The diversity of Lake Kissimmee's birds is notable, and the decline of hydrilla and the return of the natives should help the birds prosper. As we slowly fanned through the restoration area, we saw roseate spoonbills, whose flaming pink-red plumage rarely brightens this part of Florida. We also encountered another bird, the Everglade snail kite, which is an endangered species. By 1964, the U.S. population had dropped to 20, but conservation efforts brought that number up to about 1,000 by the 1990s. Most surprising to me, we saw some huge, elegant birds whose white-with-black-tip wings spanned more than seven feet as they slowly flapped overhead. Harris smiled at my incredulity and confirmed that they were indeed whooping cranes, one of the rarest birds in North America. Only about 250 live in the wild, including the flock of about 70 that recently had been introduced into the Lake Kissimmee area.

Biologists tend to use the phrase "biologically dead" when talking about melaleuca forests, which thrive in South Florida's wetlands. Walking through a stand of melaleuca trees, I felt the truth of the phrase. The woods were mute, bereft of the rustlings and buzzes and trills of a thriving forest. Even footfalls were muffled by the deep carpet of leaf litter. Eighty-foot melaleuca trees hovered above me, a scattering of one-foot melaleuca seedlings grew at my feet—and that was about it. My close inspection of the forest floor revealed only a handful of other plants representing just a few species. Without the plant life, there's not much animal life either. In melaleuca trees at the edge of the forest, birds may sometimes roost, but I didn't see or hear any within the dark interior. This melaleuca stand was largely lifeless.

My walk took place on the southwest shore of Lake Okeechobee. I'd travelled south from Lake Kissimmee about 50 miles, following the rainwater that drains from Kissimmee and the many other Central Florida lakes, flowing through the Kissimmee River, Fisheating Creek, and dozens of smaller streams down to Lake Okeechobee. *Okeechobee* means "Big Water," and big it is—at 750 square miles, it's by far the largest lake in Florida. It is the source of the water that gives life to the Everglades.

Nowadays many people equate the Everglades with Everglades National Park, but the park's 1.5 million acres encompasses only about 25 percent of the historic glades. Originally the broad, shallow river of grass—roughly 50 miles wide and six inches deep—flowed south about a hundred miles from Lake Okeechobee to Florida Bay, at the tip of the Florida peninsula. About half has been lost to agriculture and urban development, and the remaining 25 percent remains in a seminatural state in what are called Water Conservations Areas. The major WCAs lie north and east of the park and are home to most of the melaleuca. The other large infestations sprawl across portions of Big Cypress Swamp, which abuts Everglades National Park on the north and west.

I ventured into the marshes of Lake Okeechobee in the company of several people, among them Amy Ferriter, an environmental scientist with the South Florida Water Management District, Jay Hamm, a field manager for C&N Environmental Consultants, and Greg Jubinsky. Ferriter directed the Lake Okeechobee melaleuca control project from its inception in 1993 until 1996, and Hamm oversees the field crews that engage in hand-to-hand combat with the leafy beast. Hamm picked us up in an air boat and whisked us out into the marshes to see his crew's handiwork. On the way we passed a thick stand of the invader, the first melaleuca I'd seen up close.

Melaleuca quinquenervia is an evergreen from the Australia-Malaysia region. In appearance it resembles its Australian cousins, the eucalypts. Its thin, multilayered, peeling bark gave it the popular name "paperbark tree," but in recent years the Florida scientific community has talked about this menacing invader so much that most residents know it as melaleuca. Many biologists consider it the most serious exotic-species threat to the Everglades and the other wetlands of South Florida. If it isn't adequately controlled, melaleuca probably will take over most of the region's natural areas. Already it is present in about 450,000 acres, some 50,000 of which are melaleuca monocultures, and it has been rapidly expanding. One estimate pegged its spread at 50 acres a day, though Ferriter thinks this is too high now that control efforts have revved up. But if unchecked, melaleuca could invade virtually all of South Florida's wetlands within 50 years.

Hamm slowed the air boat as we entered the thickly vegetated marshes that occupy about 100,000 acres of the lake. Vast as it is, Lake Okeechobee is exceedingly shallow. You can boat into some parts of its midsection, out of sight of land, and stand on the bottom, your head and shoulders above the surface. However, even a few feet of water is too much for melaleuca. It prefers just a few inches of water or seasonally wet habitat, so it sticks to the marshes or the banks.

At little more than a brisk walking pace, Hamm twisted and turned our boat along the narrow path through the marsh. Now and then we had to duck where the lush vegetation arched across our trail. I began to admire this lovely waterscape, but before I let my admiration ripen, I asked Ferriter how many of the diverse plant species engulfing us were natives. She replied that 99 percent were native. Then I felt free to wholeheartedly enjoy the generous white flowers of the water lilies, the tricolored herons roosting on shrubs, the aquatic grasses blading into the air several feet higher than our heads, and the osprey that hovered 50 feet above the marsh scanning the fecund waters for a fish dinner. Quite a contrast to the barren melaleuca forest, which had replaced marsh that had looked much like this.

Now and then we passed the skeleton of an isolated melaleuca tree; to prevent spread, Hamm and his crews had killed these outliers first. As we motored closer to a dense melaleuca stand, we passed more and more skeleton trees. We were traveling from the edges of the infestation in toward its heart, just as the control crews do. Presently the verdant native vegetation thinned and we emerged into a fairly open area peppered with live melaleuca, about a hundred yards from the edge of the mature melaleuca forest. Melaleuca trees grow three to six feet a year, so it doesn't take them long to reach their adult stature of 80 to 100 feet. Our site was a recent colony in which the young melaleuca ranged in height from about 3 to 20 feet.

One of the crews recently had been working here—two crews of eight people, laboring year-round killing melaleuca. It's tough work. This time of year, during the summer rains, the crew spends much of its time slogging through waist-deep water, assaulted by mosquitoes and the heat and humidity. Hamm pointed out an area thick with little melaleuca seedlings and estimated that the crew would encounter about 10,000 melaleuca plants in an area the size of a tennis court.

Ferriter added, "The density here is mild compared to what you'll find up the way."

The crews pull the small melaleuca seedlings. For trees in the 10-to-20-foot range, the crew resorts to the "hack and squirt" method. With a machete, they hack down the adolescent trees and then they whip out a spray bottle of herbicide and squirt the stump. Just cutting them down is worse than pointless. "That just pisses them off," said Jubinsky. He said that cutting a melaleuca stimulates it to shoot up new stems and to drop its seeds—and, like many invasive plants, these fertile plants are prolific seed producers. A single mature melaleuca tree harbors from 2 to 20 million seeds. When crews get to trees that are too big to chop down with a machete, they girdle them and apply herbicide in the wound. Under some circumstances, managers can spray dense infestations effectively from the air, but the ground crews must do most of the work, which makes melaleuca control a laborious, expensive process.

Ferriter and other managers think that these painstaking control efforts are substantially slowing the melaleuca stampede, though there's not nearly enough funding to oppose melaleuca on all fronts. People continue to search for easier, cheaper, chemical-free methods of control. People like Willey Durden, a technician with the Agricultural Research Service.

On a July afternoon in 1997, Durden stood beside me and a photographer from a Naples, Florida, newspaper on the helipad at Big Cypress National Preserve. We sweltered in heavy, fire-resistant flight suits, waiting for the incoming helicopter to touch down. As soon as the aircraft settled onto the asphalt, the pilot waved us over and we ducked into the four-person compartment. In various ways, it was important for each of us to get out to the helicopter's destination, but the most important passengers, the real VIPs, rested in Durden's lap. He had the snout beetles. This species comes from melaleuca's Australian homeland, and they were born to eat these alien invaders.

The helicopter was hauling us and the bugs to a remote site on this National Park Service property. Big Cypress National Preserve consists of 729,000 acres of Big Cypress Swamp—about half of the original area, much of which has collapsed beneath Florida's development rush. Big Cypress Swamp is not really a swamp but a blend of mixed hardwood hammocks (tree islands), marshes, mangrove forests, sandy islands marked by slash pine, dry prairies, and vast wet prairies dominated by saw grass. Though loggers long ago took almost all the mammoth bald-cypress trees, other cypress species cover about a third of the preserve. Bromeliads and orchids bloom in the trees and black bear, mink, deer, river otter, alligator, and other animals roam the land and waters. This nearly trackless backwater even shelters the Florida panther and red-cockaded woodpecker, two of the nation's rare and endangered species. Unfortunately melaleuca has moved in and now occupies, at varying densities, about 100,000 acres of the preserve.

We whirred northward, our tiny helicopter shadow bending across a saw grass prairie veined by cypress-lined channels of water and dotted by occasional cypress domes—round, bubblelike clumps of cypress trees. Not surprisingly, the landscape looked much like parts of Everglades National Park, which border the preserve to the east and south. After about 15 minutes, we spiraled down to a landing on the open prairie. We shucked our flight suits and helmets before stepping out of the helicopter into several inches of water and mud. The copter lifted off and we squelched toward a stand of melaleuca, where several scientists and preserve staffers stood waiting for us—well, actually, they were waiting for the bugs.

As we slogged the 200 yards to where the others waited in the shade of the melaleuca, the water level gradually dropped. By the time we entered the grove, we'd left the standing water behind, though the ground was soggy. I remarked on this change to two members of the waiting group, Greg Jubinsky and Tony Pernas, a preserve manager. Jubinsky replied that I'd just happened onto one of the two main reasons melaleuca threatens to overrun South Florida.

Much of South Florida is flat. Not flat in the undulating way that the Great Plains or a desert basin is flat but truly as flat as a board. The land slopes to the Gulf of Mexico at about two inches per mile.

Mountains? Not a chance. Hills? Not a one. Jubinsky says that the highest point in the region is a landfill, and I don't think he was kidding. However, the main road through Everglades National Park does climb over one ridge. A sign marks the pass: "Rock Reef Pass. Elevation 3 feet." Chains aren't required. Whatever the slope, physics still compels a sheet of water to flow downhill; but in Big Cypress, water takes a whole day to travel maybe half a mile. In other places the flow is even slower. After the rains ease in October, the glacial pace of drainage extends the wet season, which begins in May, by two or three months.

Melaleuca doesn't mind getting its feet wet some of the time; in fact, it thrives in moist, seasonally flooded habitats. Yet it prefers somewhat drier conditions in which to begin its aggressive campaign—to set seed and germinate—so the extra-long flood season of the flatlands could cramp melaleuca's style. But no such luck. What the lack of topographical relief gives melaleuca-fighters with one hand, it takes away with the other: South Florida's flat topography allows melaleuca to change the habitat to its liking. Melaleuca trees drop a large amount of leaf litter and bark. That debris accumulates and eventually raises the elevation of the land by an inch or two, and in this terrain that's enough to change the site's hydrological cycle to suit melaleuca's reproductive needs. Melaleuca also transpires profusely, exhaling more water vapor than most native plants, which further dries the ground below the trees.

Fire is the other major mechanism that fuels melaleuca's advance. It's counterintuitive, given the watery nature of the region, but wildfire plays a vital role in most South Florida ecosystems, including Big Cypress. "This is a fire community," said Pernas. "Cypress and pine are fire-adapted, but they pale in comparison to melaleuca. The volatile oil in melaleuca leaves causes the fire to burn so hot that the heat kills the cypress and pines. But melaleuca will not die. It's incredibly fire-resistant." In a cycle much like that of cheatgrass, melaleuca's density increases as cypress and pine populations diminish with every blaze—until finally there's little left but the Aussie import. "You may end up with two or three species where you had a hundred before," said

Pernas. "The white-tailed deer rely on the natural vegetation for food, and they don't eat melaleuca. If you lose the natural vegetation, you lose the deer. Since deer are the primary food for the panther, if the deer population declines, eventually you lose the panther, too. It's all related." He added, "The red-cockaded woodpecker relies on these pine forests. If you lose the pines, you lose the woodpeckers. They'll die off before they'll use melaleuca as a nesting site."

Australian snout beetles, on the other hand, will die if they don't go to melaleuca. The plastic container in which Durden carried the bugs was filled with melaleuca cuttings. But the time had come to transfer these potential saviors onto living trees. Their release was the culmination of years of searching and screening, of a quest to find insects that would feed on melaleuca in a way that caused serious harm but that wouldn't feed on any native plants or crops. Ted Center, an entomologist and biocontrol expert with the U.S. Department of Agriculture, said that eventually he and his colleagues might release as many as ten different species of melaleuca-bashing bugs in order to hit the alien trees from many angles at once. Resource managers would continue to simultaneously assault melaleuca with herbicides and machetes. Biocontrol is part of their integrated pest management approach.

Center has been leading the antimelaleuca biocontrol program, so it was fitting that he led our group that day in Big Cypress. He dressed appropriately for the big occasion, wearing jeans and a T-shirt that sported the image of an insect. When the group pried the lid off the container that held the beetles, Jubinsky said "Where's the champagne, Ted? I don't see the cooler." Indeed the mood was festive as everyone took a share of the five dozen beetles and set out to find them a new home.

Center told us to find young, flowering trees and place the bugs directly on the savory new growth: "They've got to feed for a while on the right kind of vegetation before moving on." Grinning, Pernas came over and told me that "within seconds one that I released was drilling a hole in a leaf." Jubinsky carefully set a couple of snout beetles onto a likely branch and left them with a biblical directive, grandly intoning, "Go forth and multiply!" Exactly. Center told me that he expects the bugs to multiply from dozens to millions. He added that this species of beetle disperses as far as ten miles from release sites, and that he and his colleagues will soon be unleashing Australian snout beetles at other places around South Florida.

Melaleuca doesn't grow nearly as densely in Australia as it does in the U.S., due to the presence of coevolved predators and competitors. The tree must have seemed innocuous when John Gifford, a University of Miami Forestry professor, requested and received melaleuca seeds from Australia in 1906. He was looking for a tree that would prosper in South Florida. Seedlings were planted at his home in the Miami area and nearby at the Stirling Nursery. Independent of Gifford, in 1912 A. H. Andrews, who owned a nursery just northwest of Big Cypress Swamp, also imported melaleuca seeds from Australia. The two nurseries sold melaleuca to people looking for fast-growing trees as ornamentals, as windbreaks, and for fence rows. When people realized that melaleuca transpired water prodigiously, they conceived another use for it: drying out those pesky wetlands that interfered with development. In 1936, H. Stirling, of the Stirling Nursery, even dumped melaleuca seeds over the Everglades from an airplane.

No one knows why melaleuca didn't overtake the Everglades 50 years ago. It would seem that given its early introduction and boost from humans, melaleuca would have emerged as a major invader long before it did. Melaleuca and many other alien invaders remain quiet for long periods and then erupt. Biologists often refer to this phenomenon as *lag time*. Though the concept is poorly understood, it is crucial because lags can lead people to underestimate and accept non-native species that seem benign but later become malignant.

There is always some delay between the arrival of an invasive exotic and its emergence as a problem—even the most virulent species can't cause significant harm when populations are as small as a handful of specimens. True lags are biological in nature, not merely delays that result from the failure of inattentive human beings to notice an invasion.

Jeff Crooks, a researcher at the Scripps Institution of Oceanography in La Jolla, California, and Michael Soulé, professor emeritus at the University of California, Santa Cruz, have focused on exotics for some time. In addition to inherent traits, such as reproductive and mortality rates and dispersal strategies, which influence lag times for all species, Crooks and Soulé see two broad categories of factors that incite a long-dormant population to grow explosively: genetics and the environment.

Crooks and Soulé call genetics "the great unknown." They know of only one case in which scientists can point precisely to a genetic change that altered a species in a way that allowed the population to run rampant. The practical difficulties of pinning down such genetic changes are almost overwhelming. However, Crooks and Soulé consider it "quite likely" that such advantageous changes occur. As examples of the genetic qualities that might shift and thereby enhance invasiveness, the authors cite physiological vigor, metabolism, growth rate, developmental patterns, resistance to toxins, and vulnerability to predators or herbivores.

In a 1996 article published in the proceedings of the United Nations Conference on Alien Species, Crooks and Soulé note that population size greatly affects the likelihood that advantageous genetic changes will occur and lead to improvements in evolutionary fitness. Small populations (fewer than a hundred individuals) rarely evolve in ways that help them expand. On the contrary, the low genetic diversity of such tiny groups usually reduces their fitness. Only when individuals number in the thousands are advantageous changes likely to be incorporated into the population and harmful changes removed. "What this implies," write Crooks and Soulé, "is a positive feedback between population size and the chances that the population will improve genetically. It also implies that the longer a population exists, at least if it numbers in the thousands, the more likely is a genetic "discovery" that enhances invasiveness. Similarly, the longer a population sticks around, the more likely it will receive fresh genetic material from repeated introductions in the same place." Resource managers and policy makers had better take note; eradicating an invader before its population climbs into the thousands is important.

Scientists have a much firmer grasp of the environmental factors that allow wallflower exotics to abruptly tango. Predicting which species will run wild under what circumstances is, however, still guesswork. *Musculista senhousia,* a small marine mussel, appeared just offshore San Diego, California, in the 1960s. By the 1980s, it commonly formed dense patches of about 10,000 individuals per square meter. In 1995, after exceptionally heavy rainfall and strong red tides altered conditions in those waters, densities of this mussel rocketed as high as 200,000 per square meter. The cut-leaved teasel came to New York in the nineteenth century, but for many decades it hardly spread. Then in the 1960s, it unexpectedly headed west, forming monocultures and displacing native plants throughout much of the Midwest. Why the sudden assault? It turns out that teasel followed the newly built interstates and the other new highways that proliferated in the second half of the century. When managers removed feral goats from Volcanoes National Park, on the island of Hawaii, the alien grasses on which the goats had grazed spread out of control.

All sorts of changes have the potential to produce conditions that stimulate the rapid population growth and expansion of exotics that have been lagging. Biologists usually call this kind of environment-altering change a disturbance. The significance of disturbance goes far beyond its influence on lag time. To a large degree disturbance, or its absence, often determines whether a non-native species can get established in the first place and how far and fast it spreads from the beginning.

Scientists used to think that, with few exceptions, exotic species could invade only disturbed ecosystems. The conventional wisdom held that invaders couldn't penetrate a healthy, diverse native community populated by its full array of species. In such a pristine place, the natives fill all the niches and use all the resources, according to the theory, and therefore invaders can't gain a toehold. Oh, a few individual non-native plants or insects or birds might find a home, but they're not going to spread and dominate, not while they're surrounded by a robust native community.

In recent years, scientists have modified their initial thinking on disturbance. They discovered that more than a few exotic species possess the ability to shoulder their way into intact native communities. Successful invasion depends not only on the condition of the community being invaded, but on the character of the invader, whether it possesses decisive advantages vis-à-vis the natives in the system it encounters. The essence of the initial thinking, however, was accurate; disturbance does pave the way for invasion. Even those species that don't require disturbance usually find it much easier to enter and dominate a disturbed community.

Sadly, research on the ability of pristine ecosystems to resist invasion has no precise relevance for the vast majority of the American landscape. Almost the entire nation, except Alaska, has been disturbed in ways that encourage invasion. About 95 percent of the forests in the lower 48 have been logged at least once; most have been clearcut repeatedly. Nearly 400 million of the 2 billion acres that constitute the U.S. (excluding Alaska) are disturbed by agriculture. More than 100,000 dams have drastically altered the nature of the nation's rivers and creeks. Half of our wetlands have been drained. Livestock graze on 660 million acres, including a large majority of the West. The natural rhythms of fires and floods have been disrupted. And don't forget to add the degrading effects of cities, suburbs, towns, rural houses, highways, farm roads, logging roads, trails, mines, oil and gas wells, resorts, off-road-vehicle areas, marinas, military bases, and all the other human development that has altered natural ecosystems. Consider, too, the general stress on native species and communities caused by air pollution, acid rain, pesticides, global warming, and water pollution. The United States has been highly disturbed. That's one of the main reasons it has been highly invaded.

Humans are by far the most significant source of disturbance, but natural events, too, can alter ecosystems in ways that promote alien invasion. A gopher mound can provide a tiny opening for a seed blowing on the wind, or a volcanic eruption can create an enormous opening that nearby exotics may colonize. In Florida when one thinks of natural disturbances, one first thinks of hurricanes. On average, a given piece of South Florida feels the effects of a hurricane about once a decade. Now and then that site will suffer a direct hit and get disturbed in a big way.

The last direct hit on the Everglades occurred on August 24, 1992, when Hurricane Andrew struck South Florida. While causing some $20 billion in property damage, the sustained winds of 135 to 140 miles per hour also destroyed zoos, pet stores, and fish farms, releasing all manner of non-native beasts into the wild. There was even a mass escape of primates—some 500 macaque monkeys and 20 baboons. Andrew also ripped up a lot of habitat in Everglades National Park and Big Cypress National Preserve, opening holes in the protective blanket of native vegetation and stressing native plants. The hurricane also dispersed exotic seeds and plant material, some of which no doubt found those holes. Officials braced for a subsequent upturn in the spread of alien species. In particular, Everglades National Park managers thought Andrew would set back their melaleuca control program.

Reality, as so often happens, has turned out to be more complicated. Numerous scientists leaped at the opportunity and initiated studies to see the ways in which Andrew affected exotic species in the Everglades. Carol Horvitz, a biologist at the University of Miami and one of the leading post-Andrew researchers, summarized the first two to three years of findings in a chapter in *Strangers in Paradise*. She reports that, contrary to what managers feared, Andrew apparently did not significantly spread melaleuca or two other alien tree species that rank high on the park's most unwanted list. On the other hand, ongoing examination of several hardwood hammocks showed a rapid post-hurricane growth of another exotic species—papaya. Back to the other side of the ledger, Andrew's apocalyptic winds blew down nearly all of the Australian pines at Bill Baggs State Park, enabling park managers to remove the unwanted, non-indigenous pines and replace them with native species. Then again, in some hardwood hammocks, populations of an alien fire-ant species boomed in the wake of Andrew, apparently due to the sudden abundance of pioneer plants that provided the hollow stems in which these fire ants like to nest.

This jumble of results so far has defied easy conclusions, but Horvitz did discern an important principle: "Some native organisms may be more resilient to [natural] disturbance than certain non-indigenous species." After all, Everglades natives evolved with hurricanes, so they readily recover from them; some native species even require the occasional blowdown in order to thrive. An exotic species, too, may possess traits that allow it to prosper following a hurricane or other natural disturbance, but it may not; it didn't evolve there. In contrast, human-centered disturbances almost always favor invasive exotics because many of these opportunists evolved to follow in our wake.

Did human-caused disturbance contribute to the South Florida melaleuca surge in the 1960s? Certainly, though it's hard to say just how much and in exactly what ways. Common sense suggests that the rampant logging of cypress and pine trees helped make room for melaleuca. Abandoned agricultural fields in former wetlands have proven attractive to melaleuca. But most scientists think that the key disturbance that benefits melaleuca and many other aquatic and wetland invaders involves the defining element of their habitat: water.

When we left the site where Hamm's crew had been killing melaleuca, he steered our air boat back through the marshy maze and out into the canal that borders Lake Okeechobee. As we sped up the canal we passed stand after stand of *Melaleuca quinquenervia,* some of it green, much of it dead-brown; Hamm's crews had been busy up here, too. It would be reasonable to assume that the Army Corps of Engineers made its most important contribution to the spread of melaleuca when it planted hundreds of thousands of melaleuca seedlings in Lake Okeechobee in the 1930s and 1940s—the ancestors of the trees we were blowing past, perhaps even some of those very trees. Reasonable but not strictly accurate. The Corps planted the melaleuca in order to shield from wind and wave erosion the levees it was building around Okeechobee. It is those levees that cause some of the biggest problems—those levees and the many other levees and the pipes and the dikes and the pumping stations and the 1,400 miles of canals that plumb the Florida peninsula south of the lake. Eighty percent of this vast network was federally funded and built by the Corps.

Historically the Everglades began at the south end of Lake Okeechobee. There was no defined boundary. Water from the shallow lake simply spilled slowly down the ever-so-gradual incline south toward the Gulf of Mexico, overflowing like a plugged bathtub. The jungly marsh vegetation of the lake imperceptibly melted into the wet saw-grass prairies of the glades. But some regarded the Everglades not as a natural treasure but as an impediment to economic growth. Developers wanted flood control to protect farms and towns; they wanted water to quench the thirst of the burgeoning coastal cities; they wanted to drain the glades and create farmland; and they wanted irrigation water for those farms. And they got what they wanted.

Draining, ditching, and diking started early in the 20th century and accelerated during the 1930s, but it was during the 1940s that South Florida's development fever prompted radical changes in the Everglades. During World War II, the Corps was busy fulfilling its military role. But the end of the war freed the Corps to return to its peacetime pursuits, notably civilizing the country's natural waterways, which were perceived as entirely too unruly. Politicians directed the Corps to include the Everglades among its targets, and by 1947 the Corps had mapped out the wholesale transformation of the river of grass.

The levee that we were passing in Hamm's air boat epitomizes that transformation. It blocks the overflow from Lake Okeechobee that once nourished the Everglades and diverts most of it to cities and farms and for flood control. The volume of water flowing to the Everglades has been reduced to about one-fifth its natural level, and generally that pittance of water doesn't follow the timing of natural flows. Dan Thayer heads the Vegetation Management Division of the South Florida Water Management District (SFWMD), which controls that extensive water complex south of Lake Okeechobee. Thayer says the water management project was designed so that in the wet season, when the Everglades should be getting wetter, SFWMD drains water out of them for flood control. In the dry season, Thayer continues, when pools in the Everglades should be shrinking, concentrating resources on which wading birds and other native species rely, SFWMD floods the glades, storing water for cities and farms. This is the main reason that populations of the Everglades's famous wood storks, roseate spoonbills, egrets, and herons have plunged. Surveys in the 1980s showed that the number of wading birds nesting in the glades had dropped to about 5 percent of normal. Thayer points out that altering the water cycle also alters the fire cycle. By flooding the glades during the dry season, when fires naturally would occur, SFWMD dampens this vital ecological process.

The drastic changes in the water and fire regimes weaken native organisms, says Thayer, making them more vulnerable to invasion. "Look at the distribution of melaleuca," he said. "It first shows up in the areas where the glades have dried out some, like on the north end. The plants are stressed. They're converting from saw-grass prairie to willow, myrtle, cattail, some woody species that don't really belong in the Everglades. Plants like melaleuca come in and get a foothold and then spread from there."

A long lobbying effort by conservationists in and out of government has spurred a serious effort to restore some water to the wetlands of South Florida. A multi-agency study is being conducted to determine the ways in which water can be managed for both human and environmental needs, though squabbles over who will pay have slowed the process. As part of this deal, the Army Corps of Engineers and other plumbers are undoing some of what they did. For starters, they're looking at the Kissimmee River, historically a meandering, 103-mile waterway whose seasonal floods maintained a vast complex of wetlands. But on occasion those floods also inundated the farms and towns that had moved into the area, so in the 1960s the Corps made the Kissimmee behave, bulldozing it into a 56-mile canal renamed C-38.

Under pressure, the SFWMD now is restoring some of the wetlands and floods. They're even putting some curves back in the Kissimmee. In general, however, the restoration of the Everglades has been moving more slowly than water in the Big Cypress Preserve. None of the steps taken so far has significantly eased the crush of exotic invaders that is degrading the river of grass. Even if restoration proceeded without restraint tomorrow, the Everglades' considerable problems with exotics would be far from over.

"That's a cow-horn orchid," said David Jones, the supervisory botanist at Everglades National Park. He mentioned that this pretty orchid, the area's largest and most arresting, is on Florida's threatened-and-endangered species list. Surprisingly, at least to me, we came across a cactus; I didn't associate cacti with this wet, steamy realm. This species, the dildoe cactus, is another member of the Florida threatened and endangered club. Jones pointed to some trees above us whose branches supported bromeliads, relatives of the pineapple that have severed their ties to the soil and grow on trees. Very common on these buttonwood trees, said Jones, though only about 50 species of orchid and bromeliad reside in the park—not much diversity by tropical standards. He told me not to overlook the buttonwoods, which are a common native tree species in many of the park's coastal hammocks, including Coot Bay Hammock, the one through which Jones, Greg Jubinsky, and I were walking.

If someone had spotted us as we rambled through the undergrowth, he might have thought we were lost beekeepers. Jones had issued us mosquito jackets, which include a built-in head net—de rigueur for spending time outdoors in the glades during summer. The lightweight nylon jackets are baggy and feature mesh vents designed to keep the mozzies at bay and keep you cool at the same time. My jacket worked okay, very relatively speaking; I only sweated a few quarts and I escaped with somewhat fewer than 200 bites. Jones, on the other hand, truly suffered. Being the host and a gentleman, he took the jacket with the head net that wasn't properly attached. As thousands of those little bloodsuckers swarmed over us every step of the way, inevitably some slipped under his net and worked him over. By the time we got back inside the car and removed our jackets, his face was red and swollen from countless bites. He looked like a boxer at the end of a losing bout. I began to understand the historical reputation of the Everglades as a place fit only for alligators and fleeing criminals.

Mosquitoes and muggy 95-degree heat notwithstanding, we had parked near the end of the park's lone paved road, near the southern end of the land portion of the Everglades, in order to explore a small piece of this raw wilderness that somehow encompasses both heaven and hell. We had dived off the road into the green, first slogging through shin-deep water in a fetid mangrove stand, stumbling over the knobby breathing roots that poke up everywhere, then hummocking across a wet coastal prairie to the edge of Coot Bay Hammock. Jones wanted to show me what was being lost to *Colubrina asiatica,* a.k.a. lather leaf. This vinelike shrub from tropical Asia has erupted in some of the park's mangrove swamps and coastal hammocks, burying the native communities in a leafy green coffin.

As we ducked farther into Coot Bay Hammock, Jones suddenly called out a warning. Pointing to a small tree in front of us whose leaves were halved by a distinctive yellow midrib, Jones identified the species as manchineel. "This is the mother of all toxic plants," he said. "It's much more potent than even poison ivy. Very dangerous, very toxic." He said its poisons could leave a horrible rash like a burn for weeks. "You want to stay away from it." We did, skirting that tree and the many other manchineels that we subsequently encountered as we moved deeper into the hammock. I felt no less standoffish after Jones informed me that manchineel was yet another of the plant species in the hammock listed by Florida as endangered.

Every few steps, Jones named another of the native species engulfing us. We sniffed a wild lime, its citrus-clean aroma sharp as we pinched leaves between our fingers. The white stopper tree, on the other hand, smelled like a skunked dog. Jubinsky glimpsed a small mammal, probably a marsh rabbit, hustling from our approach. We wove past palms, mahogany, and wild cinnamon, a member of the coffee family. We watched a butterfly samba by, its flutters much more shallow and rapid than those of a typical butterfly navigating an open field. We wondered if the butterfly's movements were an adaptation to the dense vegetation. We saw a large bird fly over, but we couldn't identify it through the thick canopy. Jones said it could be an osprey or southern bald eagle, both of which commonly use coastal hammocks. Or maybe an egret, a heron, or a hawk. Jones noted that other coastal hammocks also sheltered such floral and faunal riches, riches that we lose when lather leaf wins.

Awash in sweat and mosquitoes, we headed back to the car after a scant half hour. Did people ever really live here, all summer, without air-conditioners and bug spray? Filing after Jubinsky, I estimated that 500 mosquitoes clung hungrily to the back of his jacket and his pant legs. After about ten minutes, we ended up back in the core of the hammock; we'd completed a nice round circle. We looked at each other, then looked around. Everything looked the same. No mountains or 200-foot trees to use as landmarks. No orange disk of sun in a sky grayed by heavy moisture. Not even a compass—who'd take a compass on a half-mile jaunt? So we had to postpone our escape until we heard a passing vehicle and followed the sound back to civilization.

Yes, this is hard country. But it's hard only by human standards. For the plants and animals that evolved in the Everglades, this brew of land and water is paradise. Take mosquitoes. Some 60 native mosquito species reside in the park. Not exactly the kind of biodiversity that supports a thriving ecotourism industry, but an impressive bit of diversity nonetheless. Besides, more than 600 animal species, not counting the mozzies, and about 900 plant species make the 1.5 million-acre park their home. And, as Marjory Stoneman Douglas told us, there's nothing else in the world quite like the Everglades. For these reasons Everglades National Park is the only park in the Western Hemisphere to be named an International Biosphere Reserve, a

World Heritage Site, and a wetland of international importance.

As we drove back up the road to park headquarters and the main entrance, we passed mangroves, coastal prairies, lakes, bays, pine rocklands, more hardwood hammocks, and other types of Everglades ecosystems: the foundation of the park's biodiversity. Remarkable as all this diversity is, however, it's a little less remarkable when one realizes that much of that biodiversity consists of alien species. Approximately a quarter of those 900 plants are non-native. As we drove along, Jones mentioned some of the species known to be invasive. Burma reed, for example, a tall Asian grass that sports showy plumes, grows in some of the park's pinelands; if allowed to spread uncontrolled, it can alter the fire regime fueling unnaturally hot fires that can even kill the young pines. Shoebutton ardisia is taking over the understory of some hardwood hammocks. Cogon grass, a major problem plant that has become established in 27 Florida counties, has shown up in some of the park's pinelands.

Though many of these exotics have remained harmless so far, more than enough have made trouble; one researcher considers Everglades among the four U.S. national parks hardest hit by invasives. That's low praise, indeed, considering that a national survey of park staff ranked alien invaders as the most pervasive environmental problem faced by U.S. parks. The mounting onslaught from non-indigenous species is undermining the fundamental mission of the National Park Service, as stated in the Organic Act of 1916. In this founding legislation of the park service, Congress directed the service to manage the parks in order to "conserve the scenery and the natural and historic objects and the wildlife therein and to provide for the enjoyment of the same in such manner and by such means as will leave them unimpaired for the enjoyment of future generations." In keeping with this mandate, current park service policy states that exotics will be controlled "whenever such species threaten park resources . . . [and] high priority will be given to [non-native species] that have a substantial impact on park resources."

An admirable mandate and a responsible policy. But already

invasives have impaired Everglades National Park. If the park doesn't boost its control of exotics and its efforts to prevent the entry of more exotics, the natural bounty of the park will not be left "unimpaired for the enjoyment of future generations."

Don't blame the park service. They've been ahead of most in recognizing the threat posed by exotics and doing something about it. Nor should you blame the management and staff at Everglades. They've been leaders in the fight against alien species. Doug DeVries, a resource manager at Everglades, easily identifies the source of the problem: money. The same old problem. DeVries apologizes for sounding like a budget-seeking bureaucrat, but he asserts that the pace of the exotics problem has outrun the park's funding.

Right now most of the park's anti-alien budget is shipped to the eastern front—Miami lies just 15 miles from the park's northeastern corner—where a region of several million people and industrial-strength agriculture squeezes up against Everglades's boundary. This South Florida boom-region teems with exotic species, a reservoir of invaders that constantly spill across the border into the park. Melaleuca comes mainly from the east. So far the park has been able to eradicate infestations before they get established in the park and spread, but that degree of vigilance doesn't come cheap. Australian pine, another major invader, occupies thousands of acres in the eastern reaches of Everglades. Park managers have devoted a great deal of effort and money to killing these trees; the park's first program to control an exotic plant, begun in 1969, was aimed at Australian pine. Among other things, these trees destroy nesting habitat for endangered sea turtles and the exceedingly rare American crocodile. The east side also suffers from an enormous infestation of Brazilian pepper, an evergreen shrub that forms dense, biologically barren thickets.

Each year the park spends almost all of its resource money battling melaleuca, Australian pine, and Brazilian pepper on the eastern border. That leaves mere budget crumbs for dealing with all the other invasive species problems. Nodes of Brazilian pepper have popped up throughout the mangroves and coastal prairies in a 100,000-acre swath along the western edge of the park, but little has been done to halt its expansion. Park managers don't even have a satisfactory method for removing pepper from mangroves, so they urgently need funding for research. Minor-league species are threatening to grow into major-league problems. DeVries cites seaside mahoe and Jaragua grass as up-and-comers.

His frustration was evident as he singled out the examples of cave grass and napier grass. He said that currently these highly invasive species only occupy 10 to 20 acres. At this point a few tens of thousands of dollars probably would buy their virtual eradication, but the money just isn't there. "The difficulty we have is in choosing our battles," said DeVries. "With all these [serious] exotics invading all these different habitats, which ones do you put your resources in? We still don't have enough money to treat melaleuca, and that's considered to be the Everglades terminator. Even though exotic plant control is considered to be a high priority . . . in South Florida and for the Park Service, there still isn't enough money even for [melaleuca]." He paused, exasperated, then added: "I don't know where to go from there." When I asked him about the prognosis for dealing with exotics, he said, "To me it looks like a huge project, and with so little money available for it, I have difficulty remaining optimistic." He did sound a hopeful note as we parted, however. "When money becomes available, when there's recognition at the highest levels that this is a priority, and resources are mustered to address the problem, then it's amazing what can be accomplished."

Sometimes I wonder about the fate of Coot Bay Hammock. Will lather leaf overrun it soon? DeVries said that lather leaf is a perfect example of a species that should have been jumped on when $20,000 would have done the job. Now it would cost millions, and the price is rapidly rising because this killer of coastal hammocks is rapidly spreading. DeVries said the park's surveys are ten years old, that lather leaf probably has tripled its territory since it was last mapped. I find it easy to worry about the orchids, the egrets, and the wild lime.

Working a little harder, reminding myself about the value of biodiversity, I even manage a few moments of concern for the manchineel and the mosquitoes. They, too, belong to the Everglades, this place like no other that we're in danger of losing.

CHAPTER 8

FISH THAT GROW LIKE CHINA PIGS: COLORADO RIVER BASIN

"Look for two eyes and a wiggle."

This was the advice I received in my search for an elusive quarry in Lake Mohave. The voice of experience belonged to BLM fisheries biologist Al Doelker. I was searching for inch-long transparent creatures that appear as little more than black dots and jiggles. Night had fallen, so it was time to catch these wigglers that seem more like water than flesh-and-blood animal. Beneath the brilliant desert stars set in a moonless sky, I joined Yvonne Marlin, a 20-something volunteer biologist, and Matt Connolly, an equally young and energetic forestry tech, who were gathered at the stern of a houseboat roped to the Arizona shore of Lake Mohave. Into the shallow water, maybe five feet deep, we lowered our tube lights until they dangled a couple of feet below the surface; then we stuck our nets in close to the lights—and waited on bended knee, poised to scoop.

Spotting razorback-sucker larvae is difficult, unless you develop an eye for the wiggles. Even then, it takes a lot of eyestrain. Once netted, these ethereal beings require delicate handling. We each gently raised our scoops into the cool night air and took great care not to touch the fragile larvae or squeeze them in the folds as we quickly nudged them out of the net, into a bucket of water. After an hour, Connolly and I had captured a couple of dozen each; Marlin had hauled in about 150.

Staring down into my bucket at the flickers of life, I was looking at hope. Razorback-sucker larvae are the newborns of this endemic fish of the Colorado River Basin—and the only larvae with a future were the ones that ended up in our buckets. Once numerous throughout the basin, razorbacks started declining when dams and other water-works changed the nature of the Colorado River and its tributaries. Lake Mohave is not a real lake but a reservoir backed up by Davis Dam, which spans the Colorado from Arizona to Nevada, just north of where these states meet California. Although dams may have ensured that the razorback sucker would end up on the federal endangered species list, introduced fish will probably prevent them from ever being removed from the list.

Hundreds of species of non-native aquatic animals have become established in American waters and have joined alien aquatic plants in the U.S. invasion. Green crabs are disrupting food webs and hurting the shellfish industry on both the East and West Coasts. Fisheries managers cringe at the mere thought of whirling disease. A European parasite that has invaded 20 states, this microbe kills fish, making them whirl madly before expiring. Bullfrogs, notorious for gobbling anything that will fit in their mouths, have moved from where they belong, in the East, to where they don't belong, in the West. Rusty crayfish are wiping out native aquatic vegetation in northern lakes.

And then there's the infamous zebra mussel, which during the early 1990s momentarily put the alien invasion on America's front pages. It began in the mid-1980s, when a ship in the Great Lakes discharged a handful of these little striped mollusks along with ballast water from some foreign port. They proliferated madly, and within a few years untold numbers of them had spread throughout the Great Lakes, where they altered ecosystems and fouled piers and pipes and anything else to which they could cling. Over the years, zebra mussels's impact

on the power industry—and electricity consumers—has been especially dramatic. Many power plants rely on lake water for cooling and
other essential functions. When zebra mussels clog the utilities' water
pipes, they experience costly shutdowns and must spend exorbitant
amounts of money cleaning and redesigning their equipment. Experts
cited in the 1993 OTA report estimated a total cost to the power
industry of about three billion dollars over ten years. This is a case of
déjà vu for the industry, which had lost a similar amount of money for
similar reasons during the 1970s and 1980s when Asian-clam populations skyrocketed. By 1997, zebra mussels had spread throughout
most of the Mississippi River Basin, and they're still going strong.

With the possible exception of the zebra mussel, the most prominent of the exotic aquatic animals are fish. The OTA report counted
127 exotic fish species in the U.S., most of which already have taken
hold in the wild. "There's probably not a system in the lower 48 that
has completely intact native fish fauna, with no introductions," said
Hiram Li, a professor in the fisheries and wildlife department at
Oregon State University and an authority on exotic fish.

Goldfish appeared beyond the confines of estate ponds in the late
1600s, making them the first exotic fish known to become established
in the United States. Today goldfish inhabit every state but Alaska.
Popular as bait fish, they probably were planted by anglers. Sport-
fishing enthusiasts have brought in many exotic species, either as game
fish or as food for game fish. The arrival of European carp in the U.S.
in the nineteenth century provides a classic example. George Laycock
tells the tale in his 1966 book, *The Alien Animals*.

In 1876, the federal government sent one of its fish culturists,
Rudolph Hessel, on a mission of the utmost importance: bring back
the carp, the wonder fish of Europe. His first batch died on the rough
voyage home, but in 1877, Hessel safely escorted 345 robust carp to
ponds in Boston's Druid Hill Park. They prospered and soon crowded
their new home. No problem. Congress appropriated funds to bring
overflow carp to Babcock Lakes in Monument Park in Washington,
D.C., where they duly arrived in 1878. By now people from all over

the country clamored for carp. Members of Congress fought over the wonder fish, which politicians saw as prime pork to be proudly brought home to their districts. In 1883, 298 members divided up 260,000 carp—only three members didn't get any—and the fish were sent packing to bodies of water all over the country. Happy carp owners, such as one Samuel Johnson of Savoy, Texas, wrote letters to the U.S. Fish Commission expressing their delight: "My carp . . . are doing well. They grow like China pigs when fed with plenty of buttermilk." Ominously, Johnson reported that his carp ate everything he gave them, though they favored biscuits.

Within a few years the tone of the letters had changed. People wrote that carp didn't taste that good and that it hadn't turned out to be much of a game fish. Worse, the carp were ruining habitat that sustained real game fish and waterfowl. Typical of the response to the rapidly spreading carp was an 1895 letter from George T. Mills, the fish commissioner of Nevada: "Time has now established [the carp's] worthlessness, and our waters are suffering from their presence. As food fish they are regarded as inferior to the native chub and sucker, while their tenacity to life and everlasting hunger give them a reputation for 'stayers and feeders' unheard of in any fish reports I have seen. . . ." Reports poured in telling of carp devouring aquatic vegetation and the spawn of desirable fish and of carp stirring up mud until lakes turned brown. By the turn of the century, the U.S. Fish Commission and others started working to remove carp, but it was too late. Today the wonder fish occupies every state but Alaska. On the Nevada side of Lake Mohave, I saw throngs of carp rooting in the shallows, probably descendants of the fish that dismayed George T. Mills more than a hundred years ago.

Despite the lesson of carp, sport-fishing boosters in and out of government have continued recruiting exotic fishes. From a recreational point of view, many species have succeeded. Take the brown trout. A European import, *Salmo trutta* arrived in 1883, at the start of America's ill-fated love affair with the carp. Unlike carp, brown trout became more popular over time. Large, full of fight, and more resistant than natives to intense fishing pressure, browns now lure anglers in 47 states and rank as one of the nation's top game fishes. Such cases account for the motivation behind the fact that 44 percent of the non-native fish species in the U.S. were stocked for sport; another 16 percent also are tied to recreational fishing. And anglers have introduced non-natives by dumping live-bait fish into streams and lakes.

A small but growing percentage of alien fishes are entering and spreading in U.S. waters even without intentional help from humans. The Great Lakes are particularly vulnerable because they're linked to the Atlantic. One troublesome fish, the European ruffe, slipped into the Great Lakes in the 1980s via expelled ballast water. The Great Lakes Fishery Commission expects this finned invader to harm commercial fisheries if not controlled, projecting a cost the U.S. economy of more than $90 million annually.

The sea lamprey also shows what an unrestrained invader can do. This parasitic, eel-like fish kills other fish by clamping on with its big, vampire mouth and draining the life out of them. Somehow the lamprey got into Lake Ontario, the easternmost of the Great Lakes, but natural barriers kept it from moving to the rest of the lakes. That wall was breached in the 1920s, however, when the Welland Canal opened a path between Lake Ontario and its great brethren. During the next three decades, the lamprey slowly spread, devastating commercial and sport fisheries, most notably lake trout and the multimillion-dollar industry they supported. By the 1950s, lake trout were extinct in Lake Ontario, nearly extinct in Lake Michigan and Lake Huron, and reduced to remnant populations in Lake Superior. Still hungry, sea lamprey turned their attention to whitefish, burbot, lake herring, and various cisco species. Preyed upon by lampreys, by another exotic fish called the alewife, and by increasingly desperate fishers, these native fish populations crashed.

Eventually scientists discovered a way to control sea lamprey. Lamprey migrate upstream from the lakes to spawn in gravel beds. After hatching, the lamprey larvae live in the gravel beds for two to three years, during which time they're vulnerable to chemicals. This control effort costs about $10 million a year. However, a U.S. Fish and Wildlife Service report estimates that, if left uncontrolled, sea lamprey would cost about $500 million a year in lost fishing opportunities and indirect economic impacts.

Our group's contribution to the ongoing effort to save the razorback sucker began one warm February afternoon at the Cottonwood Cove marina, on the Nevada shore of Lake Mohave. There were eight people besides myself, most of them government fisheries professionals who were donating their off-duty time. Sometimes these work parties include dozens of people. We loaded gear onto a houseboat and two small boats and headed across the reservoir to Yuma Cove, an undeveloped site on the Arizona side. Low sandy hills held together by creosote bushes and saltbush backed our anchorage. The shoreline, like so many shorelines and stream banks in the Southwest, bristled with salt cedar, a scrubby exotic tree.

The lower Colorado River (that part south of Glen Canyon Dam, near the Arizona-Utah border) runs through a harsh expanse of desert. Like all deserts, this region harbors a low diversity of native fish species. Only a few have been able to evolve to meet such demanding conditions, which makes these fish both valuable and vulnerable. Starting with the completion of Hoover Dam in 1935, engineers altered those conditions with dams and other waterworks. Dams held back spawning gravels and nutrients. The timing and magnitude of floods changed. Dams released cold, clear water from the depths of their reservoirs, cooling and clearing the historically warm and muddy Colorado. Dams interfered with spawning migrations. The razorback sucker, the bonytail chub, the Colorado squawfish, and other native species suffered.

Those new reservoirs on the Colorado drew the promoters of industrial sport fishing the way pristine coastline draws developers. The U.S. Fish and Wildlife Service, the state fish and game departments, fishing groups, individual anglers—they all figured they knew how to improve on old-fashioned, inefficient Mother Nature. But in this case, the tinkerers didn't look to foreign lands. They found the exotics they wanted right in the U.S. of A., including rainbow trout, perhaps America's most popular game fish; largemouth bass, another crowd pleaser; flathead catfish, beloved by fishers because its size—maybe 4 feet long and 50 pounds—makes for great fish stories; and, probably most popular of all in the lower Colorado today, striped bass. Nothing like a striper for size and fight. All four are native American fish, but none is native to the lower Colorado, which makes them transplants, a subset of non-native organisms.

Really, many of these fish aren't native to anywhere, unless you count government hatcheries as homelands. Rainbow trout, for example, are a manufactured product. For decades, fisheries managers have crossbred trout stocks from all over the world in a Frankenstein-like effort to create the perfect fish for anglers. The result has been trout that survive well in hatcheries and are easy for fishers to catch. What's been lost are traits needed for survival in the wild, such as hiding from predators. When these generic rainbow trout are dumped in the Colorado basin, they're exotic times two.

The OTA study found that of the 127 non-indigenous fish species in the U.S., 70 are native U.S. fish that have moved—or, more accurately, have been moved—beyond their ranges. Large as that percentage is, it doesn't show the full magnitude of the transplanting epidemic. Government and private sport-fishing boosters have introduced a few prized species to just about every gallon of water that didn't already have them. Fisheries personnel used to fill railroad cars with favored species and dump them into rivers while the trains paused on bridges. People took pack trains of mules and horses loaded with exotic fish high into the mountains. Today trucks haul the output of hatcheries to lakes, rivers, creeks, and ponds throughout the nation. Bodies of water too remote to reach by road are stocked by dropping payloads of non-native fish from planes and helicopters. Flathead catfish now inhabit 35 states, striped bass 36 states, largemouth bass 42, and rainbow trout 48. The Colorado River Basin shows the thoroughness with which exotics have been sown. Not one significant lake, river, or creek in the entire basin, from Wyoming to the Mexican border, is free of exotic fish. Sometimes it's hard to remember that many fishers ardently support the conservation of native fishes and deplore the indiscriminate introduction and stocking of exotics.

That afternoon on Lake Mohave, a few hours before it was time to catch the razorback larvae, everyone went out in two small boats to set nets for the the full-grown razorbacks. I went with Chuck Minckley, a U.S. Fish and Wildlife Service biologist and the coordinator of the project to save native fish on the lower Colorado, and Lesley Fitzpatrick, a Fish and Wildlife biologist for the branch of the agency that deals with endangered species. When we got about 50 feet from shore, we tied a bag filled with rocks to one end of the 300-foot trammel net and dropped it over the side. Minckley slowly motored away from shore as we paid out the rest of the net. To the other end of the net, we tied a buoyant, waterproof light; we would need that marker when we returned in the middle of the night to see what we'd caught. We set three more nets in other parts of the lake and then went back to camp.

At about eleven o'clock, after we'd eaten dinner and caught the larvae, Minckley steered us out to the blinking light that marked the first net. I didn't see any other lights on the water or on the shore, though we could see the glow of Las Vegas about 40 miles to the northwest. Putting our backs into it, we began hauling in the heavy net. Presently a wriggling shape rose from the dark water into the canopy of light cast by our boat's overhead lamps. Our first razorback of the night. Fitzpatrick quickly but gently untangled the writhing fish and slid it into the water in the holding tank. We continued hauling and soon another razorback emerged in the net and joined its kin in the tank. We spent another 20 minutes bringing in the net and untangling fish. Several times we pulled in carp, which feast on razorback eggs and larvae. Anglers think of carp as trash fish and fisheries managers treat them as pests. What would Rudolph Hessel, the 19th-century fish culturist who brought carp to America with such fanfare, think of his wonder fish now?

Altogether we caught about 20 mature razorbacks in that first net. Minckley, Fitzpatrick, and Al Doelker, the BLM fisheries biologist, proceeded to the essential work of the night: gathering data for their ongoing survey of the Lake Mohave razorback population. I insinuated a long-handled net into the tank and plucked out one of the fish, which I quickly passed to Doelker—one shouldn't keep fish out of water for too long. He deftly weighed and measured it (adults average about two feet long and eight pounds) and checked to see if it was ripe for spawning. As Doelker held his subject on the measuring table, I got my first good look at this rare, imperiled, and rather weird-looking species. The logic behind its common name seemed obvious as I gazed down on the rounded, suction-cup mouth and the pronounced keel that rose abruptly from the back of its head and ran down its spine— a feature that had stabilized the razorback in the turbulent currents of the pre-dam Colorado.

When Doelker finished, he held his slippery customer steady while Minckley passed an electronic scanner over the fish to see if it was a "recap" (recaptured). Every razorback they catch gets a pit tag: a fuse-size identification device that is injected into the body cavity of the fish with a syringelike instrument. If a fish was previously caught, the scanner identifies it from the tag. That allows the biologists to compare the current weight and length of the fish with its size at some time in the past. When the biologists consolidate information from thousands of recaps, they begin to get a picture of how the razorbacks are faring. If the fish lacked a tag, Minckley injected it with one.

Later each year, Minckley and friends survey bonytail chub, the other endangered species hanging on in Lake Mohave and the lower Colorado. Minckley said that they'll net thousands of carp and get maybe one bonytail. A year's survey may yield six to ten individuals. "The bonytail chub is in worse shape than the razorback sucker," said Fitzpatrick. "Why it's not extinct I don't know. Most of what's left is sitting right out there in Lake Mohave." She thinks a few remain in the upper Colorado River Basin and a few to the south in Lake Havasu, though no one has seen a bonytail in Havasu for 10 or 15 years. In all the universe, which for the bonytail is the Colorado Basin, as few as several thousand exist, according to Fitzpatrick's rough estimate—a sorry fact, considering that hundreds of thousands once graced the river network.

The third endangered native fish, the Colorado squawfish, is endangered only in the upper basin. In the lower Colorado it's extinct. Minckley said that no one has caught one in Lake Mohave for decades. Smiling a bit wistfully, he talked about this species of minnow. Hardly our stereotype of a minnow, the squawfish was a big, sleek beauty, shaped like a torpedo. As long as a man and weighing up to a hundred pounds, the squawfish used to be the system's top predator.

Razorbacks are faring better than bonytails and squawfish, but they're in sad shape compared with a century ago, when hundreds of thousands, perhaps even millions, plied the Colorado Basin. Lake Mohave is their stronghold; 20,000 to 25,000 razorbacks inhabit the reservoir. Though such numbers are alarmingly small by historical standards, they probably sound pretty good to someone accustomed to thinking about endangered species like wolves and grizzly bears. But fish populations, which by nature are far larger than those of wolves and bears, can't be measured by the same yardstick used for large mammals. Besides, it's not the size of the razorback population that most concerns biologists, but the composition. All the razorbacks are old.

Back in the 1940s, researchers began noticing the scarcity of young razorbacks. Even in later decades as more surveys were carried out, scientists found no young native fish. Between 1963 and 1990, no young survived to adulthood in lower basin reservoirs. We were witness to this eerie reality that night on the boat as we picked one scarred senior citizen after another out of the nets. "They're dinosaurs," said Minckley. Fitzpatrick added, "Probably the only reason razorbacks and chub are not extinct is they live so long. It buys us time."

But time is running out. The aged razorbacks are dying off and there hasn't been a younger generation to take their place in the world. Making matters even more dire, researchers discovered an ominous phenomenon. Studies of three Colorado Basin reservoirs showed that in each case razorbacks vanished about 40 years after the completion of the dam that created the artificial lake. Davis Dam, which plugged the river to form Lake Mohave, was completed in 1954.

This phenomenon supports the idea that the engineered conditions in the Colorado harm native fish. This is true, but layers of truth exist within this politically charged idea. In the old days, fisheries managers sowed exotic fish with hardly a second thought, but the environmental awakening of the 1960s brought challenges to their tampering with nature. To justify their continued introduction and restocking of alien species, some managers resorted to a grim logic that is still popular today. They claim that dams, pollution, irrigation withdrawals, and other impacts of development have so degraded most bodies of water that native fish are no longer fit to live in them. The managers argue that it makes more sense to bring in exotic fish that are adapted to the degraded conditions, and that's what they've done.

In many cases, however, the altered environment harms native fish populations but does not annihilate them. If left alone, they could cope. Often it's the boom in exotics, which do indeed thrive in degraded habitat, that leads to the bust in natives. The mechanism may be predation, competition for food and space, transmission of disease, hybridization, or some combination thereof. Scientists studying the curious absence of razorback young in Lake Mohave examined all these possibilities, and the evidence pointed strongly to predation. Most telling are experiments in which scientists simply exclude exotic predators from otherwise typical reservoir habitat. In these sites plenty of razorback young survive and prosper. Minckley says that razorbacks haven't evolved the defenses to escape from the non-native predatory fish. "If you didn't have any introduced fish in this river, in this reservoir you'd have razorbacks and bonytails and probably squawfish."

So, the obvious answer is to get rid of the aliens, right? "It's just not going to happen," said Fitzpatrick. For one thing, explained Minckley, eradicating introduced fish is tough. "You have trouble removing them from a five-acre pond," he said, "let alone from a reservoir or a river system." But such practical restrictions are secondary to the main obstacle: the power of fishing.

Sport fishing is big business. The American Sportfishing Association reported that in 1996 recreational fishing in the U.S. produced a total economic output of $108 billion—up 36 percent from the last such study done in 1991. The study also counted more than 35 million anglers in the U.S. Much of this industry revolves around exotic fish. I saw a manifestation of our nation's gone-fishin' passion while I was driving up to Lake Mohave. Emerging from hours in the stark Mohave Desert, I was startled by the sudden blue of the Colorado River near Parker, Arizona. I headed north along the river and immediately plunged into a vacationer's paradise. Travelers come here for a variety of reasons—the fine winter weather, water skiing, houseboating—but fishing is a key part of the mix.

I passed bait shops, burger joints, and places selling acres of boats. I saw a sign for Emerald Cove Resort: "A Treasure Chest of Fun on the Colorado River." I passed boat ramp after boat ramp, where lines of pleasure craft waited their turn to enter or exit the river. Jet skis roared across the water, no doubt eliciting curses from anglers. Mazes of mobile homes, condos, real estate agencies, and marinas appeared along the road. The dunes and hills that hadn't been developed bore the crisscrossing scars of off-road vehicles.

After 15 miles of the strip development, I came to Parker Dam, behind which lies the 45-mile-long Lake Havasu reservoir, one of the sport-fishing hotspots on the lower Colorado. But it isn't hot enough, not after years of increasing fishing pressure and declining natural habitat. So the reservoir is now on the receiving end of the Lake Havasu Fisheries Improvement Partnership Program, "The largest and most comprehensive warm-water fisheries project ever undertaken in the United States," according to one of the program's brochures. The BLM, which is the lead agency, and a consortium of public and private sport-fishing interests have $28.5 million and big plans for spending it. They're building more fishing docks, more shoreline trails for anglers, more access roads, more fish-cleaning stations, and more parking areas. To provide more bottom cover and habitat diversity, the program managers are sinking a variety of structures into the lake, including 1,050 tire towers, 3,484 Fish 'N Forests (plastic trees), 54,724 catfish houses, and 67,482 bass bungalows. Being an easy day's drive from Las Vegas, Los Angeles, Phoenix, and San Diego, Lake Havasu has long been popular—it racked up 43,000 angler use days in 1989—but in a few years, when the improvement program is finished, angler use days should hit 88,000 a year. That jump in sport fishing is expected to pump an extra $3.5 million a year into the local economy.

The Endangered Species Act demands examination of the native fish species in Havasu. In their literature, the BLM and the other backers of the Havasu improvement program claim that "[the] project also presents a unique opportunity for the cooperative recovery of endangered species of native fish." While true, the original improvement program, before a native fish component was added, did nothing for native fish. In fact, part of the plan poses a threat because it will boost the populations of those native-munching game fish and will lure razorbacks and bonytails to the artificial structures, where the predatory exotics will be waiting in ambush. So Fish and Wildlife struck a deal with the sport-fish boosters; they could have their artificial structures if they put money into a program that would put 30,000 razorbacks and 30,000 bonytails into Lake Havasu.

The razorback larvae we collected from Lake Mohave travelled to a Fish and Wildlife hatchery. After about a year, the razorbacks, now two to six inches long, returned to the vicinity of Yuma Cove but not to the reservoir. The morning after our nighttime netting, I had the opportunity to see where the young razorbacks go. Most of our group walked up the shore a few hundred yards from camp to meet a Fish and Wildlife stocking boat that had come to shore. With help from maybe ten other people who had accompanied the boat, we formed a bucket brigade and passed 3,500 razorback fingerlings up the narrow beach and over the 10-foot-high berm at the back of the beach. Behind the berm, safely separate from the reservoir and its exotic predators, lay a pond of several acres. The young-adult razorbacks would remain in the pond for 6 to 10 months, by which time they would be 10 to 12 inches long. At that point, managers will return the razorbacks to the reservoir. Research has shown that natives of this size are safe from most alien jaws. The Lake Havasu native fish program is following the Lake Mohave model. Friends of the razorback and bonytail will feel much more relaxed if two large populations of these tottering species exist.

That night, when we hauled in the second trammel net, one of the razorbacks looked strikingly different from the rest: smaller, slimmer, and unblemished. It was a young razorback, probably from the first batch of fingerlings raised behind the Yuma Cove berm in 1992. Other researchers, too, have been netting a small but growing number of young razorbacks. If the larvae in the bucket represented hope, our robust young adult represented success, however limited. The presence of the first new generation in decades goes a long way toward meeting Minckley's first goal: prevent extinction. The presence of young adults and the establishment of large native-fish populations in Lake Havasu help achieve Minckley's second goal: preserve genetic diversity. "The final step in this plan is flowing rivers," said Minckley, smiling at this happy notion but vividly aware of the difficulty in making it happen. "This is really dreaming, but we could have one, two, maybe three areas of flowing streams where you could have your native fauna [and native flora]. It'd have to be a headwater reach. It couldn't be the Colorado. You could go in and take out the exotic fishes and prevent their reintroduction—and just have native fish there."

PART II: COUNTERATTACK

CHAPTER 9

THE APHIS DEFENSE

C an we prevent all invasive species from entering and becoming established in the United States? Well, the answer to that question is no. Can we prevent a great many invasive exotic species from entering and becoming established in the United States? This time the answer is yes.

I saw an attempt to answer the question of how when I visited Kahului Airport on Maui in Hawaii. For an afternoon, I joined inspectors in the Hawaii Department of Agriculture as they checked passengers and freight for invaders. Dennis Tokuoka, the supervisor, introduced me to the operation. Tokuoka is a shortish, middle-aged man who spoke quickly and moved fast. Slipping out of the baggage-claim area through a door marked Authorized Personnel Only, we dodged baggage trains and shouted to each other above the shrill of taxiing jets until we ducked into the inspectors' office. Desks, file cabinets, tables, and bookshelves jammed the small room, which serves all seven inspectors.

Tokuoka showed me the big grease board on the wall that noted to which incoming flight each inspector was assigned. Smears told of the many erasures Tokuoka and the others had made as tardy flights turned an already untenable schedule into outright chaos. He said this day was par for the course.

The inspectors must try to handle everything coming into Maui. In addition to the main terminal at the airport, they cover the harbor, where commercial ships dock; marinas, where small boats arrive; and the airport terminal that serves small planes, where movie stars and corporate bigshots arrive at all hours of the night to avoid the glare of publicity. Inspectors have grown accustomed to phone calls at 3 a.m. summoning them to meet a Lear jet.

Though these agriculture-department employees are mainly concerned with intercepting bugs and diseases that affect agriculture, they also must look for other species on the state's prohibited list, including pets and species that threaten natural areas. This creates many extra burdens. As we spoke, Tokuoka was called to take a look at some exotic fish brought in by a passenger. He said such peripheral errands can turn into three-hour marathons if the fish turn out to be rare species he can't easily identify. Tokuoka's favorite pet story involves a cat. Cats, dogs, and other animals that can carry rabies, which so far has not reached Hawaii, must submit to a long quarantine period before being released to their owners. Dread of this lengthy separation drives people to try to smuggle their pets into Hawaii. One day the inspectors received a tip that a woman on an in-bound flight was attempting to sneak in her cat. Without the tip, they never would have caught her. Feigning pregnancy, she had tranquilized her cat, strapped it around her waist, rounded it out with padding, and hidden it under a bulky sweater.

Tokuoka took off to check out the fish, so he handed me off to Andrew Kubo, a cheerful young man who has been inspecting for five years now. I climbed aboard Kubo's golf cart and we took off to meet a United flight. The inspectors had gotten the carts only after years of pleading. Tokuoka said that prior to receiving the carts, he had worn out a pair of shoes every two months racing from one part of the terminal to another. Kubo and I arrived at the door to the plane just as it opened; a flight attendant immediately handed Kubo a stack of about 300 declaration forms.

These "dec" forms are the heart of the process, and inspectors rely heavily on passengers and crew filling out these forms honestly and thoroughly. The forms ask people if they're packing such items as fresh produce, live seafood, reptiles, fungi, or protozoa. At the bottom, the form asks those who have visited farms overseas to "wash all clothing, footwear, and persons thoroughly before entering farms in Hawaii to prevent the spread of plant diseases." Perhaps the devastating papaya virus came to the Big Island with a traveler who didn't heed this request, but no one knows; the state's prevention operation hasn't been thoroughly tested for effectiveness. The form also warns that messing with dec forms or attempting to bring in prohibited items is punishable by a maximum penalty of $25,000 and/or one year in prison.

Kubo watched the passengers stream by. He stopped one man who was toting a small cooler and asked him to open it. While examining its contents, he continued eyeballing the passing stream of people in shorts and bright shirts. Kubo says that people carrying coolers are the prime target. He lets everyone else go unless he spots something obvious, such as someone carrying a sack of oranges. When the last passenger had filed by, Kubo hurried into the plane for a quick check, then we jumped back into the golf cart and zipped off to baggage claim. There Kubo flipped through the dec forms with the dexterity of a Vegas dealer, looking for items that he'd have to check before passengers could take their bags; simultaneously he watched for coolers on the conveyor belt that was taking checked bags into the terminal. After all the bags had flowed past, Kubo hustled into the terminal to find the people who had declared prohibited items and items likely to harbor pests. Anything not declared would never be caught that day; Tuesdays through Fridays, however, the inspectors get help from a beagle that is trained to smell fresh produce and meats.

As soon as he finished in the baggage area, Kubo and I got back into the cart and sped to the freight terminal, where truckers had been waiting for hours for someone to inspect their loads and release them. Typical, said Kubo. Everyone is always waiting on the inspectors. Kubo is used to dirty looks from impatient people. Impatience can escalate into anger when inspectors have to confiscate things. Kubo recalled a container of crawfish that someone in Louisiana had sent to a former Louisiana resident living on Maui. This guy had come to pick up his much anticipated treat only to find that he couldn't take it. He whined about how he was dying for some down-home cooking, but to no avail. Finally he went and got a wok, a camp stove, and spices and returned to the airport and cooked up the crawfish in the cargo area.

At the freight terminal, several truckers opened their cargoes for Kubo to examine. One contained dozens of crates filled with lettuce. Kubo opened one crate and looked to see if it was wilted or if any bugs were crawling around on top. Seeing nothing, he gave the trucker the all-clear and then hurried to the next. There he ran his eye across the top of a jumble of crates brimming with grapefruit, eggplant, asparagus, limes, spinach, and other fresh produce. After maybe 30 seconds, he gave the okay and then rushed on to the next trucker. A study indicates that such "tailgate inspections" catch less than 40 percent of incoming pests, yet Kubo had no alternative. When he finished with the truckers, we hurried back to the golf cart and went to meet another passenger plane that had just arrived.

And so it goes. A handful of grossly overworked inspectors scramble around the Maui airport trying to catch invasive species with a net that's full of holes. Tokuoka said it's not uncommon for him or his inspectors to work 30 days in a row. Anna Mae Shishido, an inspector for 20 years, said, "We don't have a slow season. Time flies and it's time to go home, and you haven't had lunch. And you still have to do your paperwork." Sometimes whole plane loads of people and freight go uninspected because no one can get to them. Shishido listed some of the basic things their operation could use, in addition to the blatant need for more inspectors: more golf carts, at least one more beagle and handler, X-ray machines, a quarantine room, a freezer for storing things, and an incinerator for destroying things. Shishido yearns to have more time to inspect more things more thoroughly. "I'd open more bags," she said, "and spend more time at freight, especially. We definitely need to look closer." Back in their office she showed me aphids, leaf miners, mint rust, and other pests they'd discovered. She wonders what they're not finding. On average one new invasive species gets established in Hawaii every 18 days. That's two million times faster than the pre-human rate.

Not every inspection operation in the United States suffers from the malnourishment that Maui's does. Then again, some places that need inspection operations have none at all. Even with ample resources, and using other means in addition to port-of-entry inspections, preventing alien species from getting in and getting established is difficult. But we can do much better than we are. We need to do much better. Virtually everyone battling exotics believes that prevention is paramount in the war on aliens. We must try to stop the arrival of exotics or immediately stamp out the next melaleuca, the next whitefly, the next leafy spurge, and the next carp. We must slow the influx of such damaging species from a gusher to a trickle. Hawaii and a handful of other states try to prevent the entry and establishment of unwanted exotic species, but for the most part that challenge falls to two federal agencies.

The U.S. Fish and Wildlife Service regulates the importation and of injurious fish and wildlife and is responsible for preventing their interstate spread. Only a small portion of this agency's budget is used for these purposes, but the service has nonetheless mounted some vigorous anti-invasive efforts, notably its campaign to keep zebra mussels from heading West. At the same time, the service continues to be a purveyor of exotic species—mostly transplants—that fishers and hunters favor. Fish and Wildlife, for example, introduces about 300,000 rainbow trout into Lake Mohave every year. These hatchery products, however, may compete with the endangered native fish—species that the service is directed to protect and restore under the Endangered Species Act .

"That's what we've done for a hundred years," said Lesley Fitzpatrick, the Fish and Wildlife biologist who has worked to protect the native fish in Lake Mohave and Lake Havasu. "We've raised fish and put them out for people to catch." Fitzpatrick notes that the mission of the agency has broadened in recent years—after all, her office helps endangered native fish—but she acknowledges that the service's historic interest in serving the sporting public still runs strong in the agency's veins. Fish and Wildlife is feeling its way toward a new equilibrium in which conservation and sporting interests are more closely balanced. Evidence can be seen right there in Lake Mohave; though the service introduced all those rainbow trout last year, it was fewer than they used to, and they put the trout in areas least likely to shelter native fish.

The Fish and Wildlife Service devotes only a small part of its resources to invasive species, but thwarting the invasion is the defining mission of the second agency that tries to prevent the entry and spread of exotics—the Animal and Plant Health Inspection Service (APHIS), an agency of the U.S. Department of Agriculture. APHIS devotes nearly every penny of its half-billion-dollar budget to stemming the tide of invaders. As concern about invasive species has grown, however, so has criticism of APHIS. A few critics accuse APHIS of doing a lousy job. A multitude of critics accuse APHIS of not doing enough.

Many people think of APHIS as the border patrol for invasive species (the ones other than those covered by Fish and Wildlife), but APHIS tackles exotics in other ways, too, as Chuck Schwalbe explained. An entomologist and a 20-year APHIS veteran, Schwalbe currently serves as the agency's associate deputy administrator for Plant Protection and Quarantine. Schwalbe acknowledges that APHIS's border-patrol function is vital and highly visible. About 2,600 APHIS inspectors check incoming cargo, passengers, and packages at 172 ports of entry throughout the United States. But in addition to this basic function, says Schwalbe, APHIS tries to prevent the entry and establishment of unwanted aliens in three other ways. In some countries, APHIS works to ensure that cargo is free of invasives before it's shipped to the U.S. Knowing that some species will slip through, APHIS maintains a network of lookouts who report any sightings of new undesirables. When that network sights a new invasive species, APHIS cooperates in efforts to control and contain it. The need for rigorous, multifaceted prevention was put simply by Bob Flanders, a senior entomologist at APHIS: "We've got a lot of scary things coming into this country."

Much of the difficulty of prevention stems from the size of the intruders and the volume of travel and trade. If inspectors had to watch out for hippos and tigers in ten ships a year, life would be easy. But APHIS must worry about millions of ships, planes, and packages that may be transporting beetle larvae, bugs the size of a comma, tiny seeds, spores, even microscopic pathogens. How do you detect such things when faced with a thousand tons of containerized cargo? Schwalbe states the obvious: Inspectors can't come close to examining everything. He also points out that inspectors don't need to examine everything in order to do a good job. In many cases, reports Schwalbe, APHIS inspectors look at less than 2 percent of the material in a given shipment, but they figure that usually brings the risk "within statistical tolerances." The goal is not to keep out a hundred percent of the unwanted organisms—"we can never assure zero risk," says Schwalbe—but to hold the number of invaders to an acceptable level. Through its early detection and eradication efforts, APHIS hopes to take care of those non-natives that do slip in.

APHIS prioritizes inspections according to numerous criteria, such as the commodity involved (Does it often carry invasive stowaways?) and the country of origin (Does it harbor many species that could become problems in the U.S.?). APHIS fears some pests so much that goods from areas inhabited by said pest are automatically subjected to preventive treatments. Shipments from Mediterranean fruit-fly territory, for example, are routinely fumigated or subjected to cold treatment.

How well does APHIS's program work? No one knows. The 1993 OTA report concluded that little research had been done to assess the effectiveness of APHIS's inspection programs. "It is unclear what level of exclusion APHIS aims for or routinely attains," the report states, "since the agency lacks performance standards for its port-inspection activities or routine evaluation of its programs." Robert Eplee, the senior research scientist for APHIS, puts it more strongly. Currently serving as director of the Oxford Plant Protection Center in Oxford, North Carolina, Eplee has been fighting invasive species since he started working for APHIS in 1965. Eplee, not one to mince words, said, "We [APHIS] don't have the slightest idea how effective, or ineffective, our inspection systems are." He says APHIS does little quality control and just assumes they're intercepting most invaders. Schwalbe disagrees, citing a number of instances in which APHIS has monitored inspections to see how much prohibited material they were catching, but such efforts seem sporadic. A 1997 General Accounting Office (GAO) report found that APHIS has begun to improve its monitoring but as yet hasn't determined the extent to which its prevention program works.

Though critics of APHIS agree that the agency needs to better assess its effectiveness, most also agree that the agency probably has done a fairly good job of catching some of the invasive species that it's trying to catch. But what about the invasive species APHIS makes no attempt to stop? "I call it 20/20 tunnel vision," said Eplee. "APHIS can see some programs and do well on them, but they're very narrowly focused." APHIS devotes almost all its resources to citrus canker, karnal bunt fungus, screwworm, gypsy moth, fruit flies, and other insect and disease species that hit farmers and ranchers in the pocketbook. It seems natural for an agency within the U.S. Department of Agriculture to stick to agricultural invaders, but APHIS has been entrusted with the task of preventing the entry and establishment of all invasive species, including species that harm natural areas.

"It's all agriculture-protection programs," said Randy Westbrooks, the national weed coordinator for APHIS and the invasive plant liaison between the departments of Interior and Agriculture. "They're important programs, but the glass is half full. APHIS's paradigm of protecting agriculture is a sin of omission. It's not that they're doing anything wrong. They're just not doing enough." APHIS's predilections wouldn't matter if another agency could step into the gap, but that's not likely. "The thing is," said Westbrooks, "APHIS has the only inspection force in the entire country . . . and that can't be replaced. You can't have environmental inspection programs at the same ports inspecting the same commodities." Westbrooks qualified his statement, acknowledging the possibility that the U.S. Fish and Wildlife

Service could greatly expand its force of about 150 inspectors and take on the responsibility for the environment, but he considers this a long-shot. In addition, APHIS currently is in the best position of any federal agency to work with private land owners in dealing with exotic species problems.

Even within the realm of agriculture, APHIS tries to protect specific industries from specific invasive species, and the agency is reluctant to expand its horizons. According to the Weed Science Society of America, APHIS largely ignores invasive plants, even though weeds cost American agriculture some $40 billion a year. Westbrooks cites lack of funding for some of this neglect. The budget cuts made by Congress during the mid-1990s decimated his already inadequate resources. "I had seven people working for me," said Westbrooks, "and now I have me." He added, "I'm sitting here with a $454,000 budget for weeds on private lands for the entire country. In one county in Oklahoma, they got $1.2 million just to deal with brush control."

Eplee believes that APHIS simply hasn't been operating under a "resource-protecting philosophy," like some other federal agencies, such as the Forest Service and the Bureau of Land Management, which have recently begun to move toward the ideal of placing the health of the land above the interests of timber companies, ranchers, and the mining industry—their historic customers.

"One of the criteria that makes something a top priority [for APHIS]," said Eplee, "is that it have the ability to creep, crawl, or fly." He added, "They love to work on flies. That's their first love." Especially fruit flies, says Eplee. While traveling in Florida in the summer of 1997, I witnessed a striking manifestation of the agency's passion for fruit flies. Sightings of Mediterranean fruit flies galvanized APHIS officials the way an alarm ignites fire-fighters. Medfly infestation could ruin a great deal of fruit and cripple the Florida citrus industry's ability to export their wares. Together with the state of Florida, APHIS immediately initiated a major eradication campaign. At night, planes sprayed malathion over the affected region, including Tampa and surrounding areas.

Such drastic measures stir up plenty of controversy.

APHIS's response to the plant species salvinia, which Eplee describes as "one of the worst water weeds in the world" is also telling. Leroy Holms, a world authority on weeds, placed this floating plant, which has not yet invaded the U.S., on his list of the ten most undesirable invasive weeds on Earth. In 1995, a land owner in coastal South Carolina discovered some plants in his pond that he couldn't identify or get rid of. Ultimately, a botanist recognized the weed as giant salvinia and contacted Eplee and Westbrooks. The mere mention of salvinia got their full attention, and their attention escalated into alarm when they learned that the infested pond had an overflow system: Heavy rains would carry the salvinia into a stream network through which it could spread throughout the lowlands of South Carolina and possibly beyond to neighboring states.

Eplee and Westbrooks got together and contacted APHIS's national office, urging APHIS to help them move quickly to eradicate the salvinia. APHIS replied that they didn't have the funding. This didn't surprise Eplee, given his view of his agency's biases: "If it had been a gypsy-moth found down there," said Eplee, "within hours they would have had all kinds of strategies in motion." He noted that weeds don't require that level of haste, but speed still was of the essence, given the threat of salvinia washing from the pond into the river system. Eplee and Westbrooks got in touch with the South Carolina APHIS office and, with various state scientists and managers who instantly realized the danger, together they eliminated that patch of salvinia posthaste. "Now, that case never raised an eyebrow in APHIS," said Eplee, "but it was probably one of the greatest achievements we have done."

Administration of the Federal Noxious Weed Act, passed by Congress in 1974, was assigned to APHIS in 1975. This law, and the Federal Seed Act, gives APHIS the authority to prevent the entry and establishment of any plant species placed on the federal noxious-weed list. Currently only foreign weeds absent from the U.S. or that have limited distribution here can be listed. Even in light of this limitation and the cumbersome listing process, many feel that APHIS has been slow to name plants to the list. Currently the list includes fewer than a hundred plants. Several years ago, Howard Singletary, then director of the Plant Industry Division of the North Carolina Department of Agriculture, calculated that at least 750 eligible plants remained unlisted. Until APHIS added melaleuca in 1992, every species on the noxious-weed list was primarily a weed of agriculture or one that causes some other direct economic loss. And according to Eplee, APHIS "fought tooth and nail" to keep melaleuca off the list. In early 1998, wetland nightshade became the second weed on the list that mainly harms natural areas.

Eplee also criticizes APHIS for foot-dragging in general. "Programmatic activity is divided into six parts," said Eplee. "The thinking, the talking, and the planning are steps one, two, and three, and you've got to go through those. But then you get into the research, the implementation, and the appraisal. APHIS is . . . lacking in the areas of research, implementation, and appraisal. If you only get halfway there, you still haven't killed [or kept out] one undesirable species."

Sometimes APHIS moves slowly even when dealing with an insect pest of agriculture, which seems contrary to agency tradition. But occasionally considerations that go beyond institutional culture come into play. Faith Campbell, the invasive-species coordinator for the Western Ancient Forest Campaign, cites the case of the Asian longhorn beetle. In 1996, the longhorn beetle showed up in New York City and a few other scattered sites around the country. This insect attacks certain species of trees, particularly maples, and could degrade millions of acres of hardwood forests. People in upstate New York and New England worry that it could seriously hurt the maple-syrup industry and the enormous tourism business that revolves around the fall foliage—flaming-red sugar maples play a leading role in nature's autumnal show. The beetle also threatens the millions of ornamental maples that line city streets and shade yards; it is in these settings that authorities first found this exotic pest.

Campbell reports that APHIS and the U.S. Forest Service moved quickly enough to eradicate the beetle. Working with state and local officials, the agency mounted a campaign to remove infested maples wherever they were found. This led to strange and sometimes controversial scenes in which government workers tromped through houses in urban places like Brooklyn, New York, to get to maples in pocket-size backyards. "They are dealing with what they are finding," said Campbell, "but they're not cutting off the source." Authorities think that beetle larvae slipped into the U.S. in wooden packing crates from China. "APHIS hasn't done what I think they should have done within weeks of the first discovery of how it's getting in," said Campbell. "They should have said, 'No more wooden crates out of Asia until the agencies of those countries convince us they have found a way to make sure that they don't have beetle grubs in them.' APHIS has the authority to do that. Schwalbe notes that APHIS has recently stepped up its inspections of wooden crates from Asia, and that APHIS has been working with Asian nations—especially China—to solve the problem at its source.

"I think this is a prime example," added Campbell, "of APHIS not taking its job sufficiently seriously to do what it's charged with doing." But why wouldn't APHIS do all it could to shut out the Asian long-horn beetle?

Campbell, and many other observers of APHIS, believe the answer lies in the ascendancy of free trade. Campbell points out that free-trade boosters in industry and government pressure APHIS to refrain from slowing international commerce even slightly.

In a paper delivered at the 1996 UN Conference on Alien Species, Peter Jenkins examined the close ties between international trade and invasions. Currently based at the Center for Wildlife Law at the University of New Mexico, Jenkins was a major contributor to the landmark 1993 OTA report on non-indigenous species. "The most important pathway of harmful exotics into the United States is neither intentional releases nor contraband brought in by international travelers," he notes. "It is unintentional importation through international trade. Exotics frequently 'stowaway' in ships, planes, trucks, shipping containers, and packing materials, or 'hitchhike' on nursery stock, fruits, vegetables, seeds, and other imports." The OTA report shows that 81 percent of the harmful new alien species known to have entered the U.S. from 1980 to 1993 through identified pathways got in via trade goods. Like Campbell, Jenkins refers to the pressure that trade advocates apply to APHIS, making it difficult for the agency to make objective, scientific decisions. "Biological considerations regarding pest risks have not necessarily outweighed economic considerations," he writes. "Ultimate decisions about exotic species risks have a political component that can vary from administration to administration, depending on the value placed on 'free trade' versus the value placed on preventing biological damage," he adds.

The case of Siberian timber, detailed in the OTA report, illustrates the danger posed by heedless trade. In the late 1980s, after loggers had removed most every available tree in their region, some West Coast sawmills began eyeing Siberia, which contains nearly half of the world's remaining softwood timber. Though raw logs have been the pathway into the United States for a number of virulent invasive species, importers arranged for a shipment of unsawn Siberian timber to arrive at the port in Eureka, California, in 1990. The importers duly informed APHIS and the California Department of Agriculture. Worried California officials sought delays in order to thoroughly evaluate the huge risks, but APHIS allowed the shipment to proceed. The logs were fumigated and taken to the mills.

The shipment contained logs representing four species of Siberian trees. Dead insects were found on logs of three species and nematodes were found on the fourth. Two of the species were identified as potentially destructive new pests. One was the Asian gypsy moth, even more feared than the dreaded European gypsy moth. Both species defoliate forests. However, European gypsy moths normally will not eat conifers and the females don't fly; Asian gypsy moths eat both conifers and deciduous trees and the females do fly. According to the OTA report, the deputy director of the Washington Department of Agriculture said that the Asian gypsy moth "has the potential to be the most serious exotic insect ever to enter the U.S."

Following the discovery of these pests, the importers voluntarily postponed further imports of Siberian timber until APHIS could evaluate the risks. APHIS entered into discussions with the importers, deciding the terms under which massive shipments of Siberian logs would be brought to the mills in the Eureka area. Alarmed officials in California and Oregon sought outside opinions from university scientists, several of whom warned that importing these logs likely would introduce some terrible pests, even if the timber was fumigated. The scientists and the state officials eventually convinced three members of Congress from Oregon to ask the Secretary of Agriculture to restrain APHIS, and APHIS has arranged another postponement of log shipments.

A U.S. Forest Service/APHIS task force undertook a thorough study of the risks in 1991, and later that year concluded that importing timber from Siberia was a bad idea. On the upside, the logs would supply cheap raw material to a few mills. On the downside, many of North America's forests would likely be devastated, particularly western forests that have seen few exotic pests. The task force estimated that if the Asian gypsy moth and another exotic moth invaded Northwest forests, the losses between 1990 and 2040 would range from $35 billion to $58 billion.

The Asian gypsy moth did get into the U.S. but not via log shipments. While the APHIS task force was examining the impact of log imports, the moth slipped into the Northwest aboard grain ships. Federal and state agencies spent tens of millions of dollars in what they hope was a successful effort to eradicate this alien invader. Whether any Asian gypsy moths survived remains uncertain. Attempts to obtain logs from Siberia continue, and importers have expanded their efforts to include Chile and New Zealand. And they're already bringing logs from Mexico to southern Oregon.

To some extent APHIS has little choice but to stay out of free trade's way. The General Agreement on Tariffs and Trade (GATT), the North

American Free Trade Agreement (NAFTA), the International Plant Protection Convention (IPPC), and other agreements that ease international trade do not encourage the regulation of invasive organisms. "These treaties," said Westbrooks, "GATT and NAFTA and IPPC, they worry me." Speaking of IPPC, he said, "Things are getting put in there that are seriously going to limit our ability to prevent the introduction of a whole bunch of stuff that's going to be a threat. This is all happening in the name of facilitating trade." Eplee and other observers share many of Westbrooks's concerns, but they feel that for the most part the treaties don't have to be disastrous. Much depends on how officials interpret these often broadly worded pacts—which in turn depends on officials and their constituents recognizing the gravity of the alien invasion.

As an example of the treaty provisions that concern him, Westbrooks cites a piece added to IPPC during its 1997 revision. It states that a signatory nation can't regulate imports suspected of carrying invasive species that already inhabit that country unless said nation has an official control program for that species. Westbrooks agrees with the basic intent of the provision, which seeks to stop countries from unfairly shielding domestic industries from foreign competition under the guise of preventing the entry of invasive species. Westbrooks notes, however, that there are hundreds of bona fide invasive species that we aren't controlling, usually because managers don't have enough money or because they don't yet know how to control the exotic. That doesn't mean that we want more of these species coming in.

Westbrooks also is concerned about the new IPPC provision that inhibits our ability to regulate the entry of species that already are widespread in the U.S. For one thing, unless the exotic in question already has spread throughout its potential range, the new arrivals may end up in locations not yet infested by that species. For another, the new arrivals may include different biotypes: individual organisms that have slightly atypical genetic material.

Westbrooks refers to a familiar enemy, hydrilla. Already present in 20 states, hydrilla may qualify as "widespread" under the new provision. Still, Westbrooks would hate to see more hydrilla enter the country; currently, only two biotypes are present in the U.S. "But there are at least 33 biotypes of this thing," said Westbrooks, "and we definitely don't need any more of those biotypes to come in from other countries to increase the genetic diversity." New biotypes can create problems by hybridizing with the existing biotypes, injecting fresh genes into the mix. "You could get something worse than what you've got," said Westbrooks. "Then who knows what we'll have in 50 years in terms of hydrilla. Maybe some much worse variety than what we have now."

International trade agreements can undermine the regulatory capabilities of APHIS in subtle ways. Much less subtle is the fact that the government, with ample prodding from industry, has ordered APHIS to facilitate trade. "The agency over the last several years has assumed the role of facilitating trade," said Eplee, "which means they want commodities to get through the ports as freely as possible. That means that with things coming into the United States, we're not looking to keep out everything that ought to be kept out." Schwalbe acknowledges the potential conflict. "It's a question that plagues some of our employees," he said. "They look at this and say, 'I feel tension here between my mandate to prevent pests from getting into the United States and my other mandate to facilitate trade. I feel like there's some compromising going on here.'"

But Schwalbe insists that APHIS resists the pressure to compromise. He says that facilitating trade doesn't mean going light on the inspection of Taiwan's exports into the U.S. so that U.S. growers can export citrus to Taiwan. "We're not a marketing organization," he said. "We're a scientific organization." To illustrate the ways in which APHIS facilitates trade, Schwalbe sketched a hypothetical scenario involving the U.S. citrus industry. If growers wanted to ship their fruit to a citrus-growing country but that country refused entry fearing citrus canker, APHIS would explain that citrus canker only exists in a few places in the U.S., and would assure them that the citrus shipped to their country would be certified as clean.

This scenario evokes a new function that APHIS has taken on in the name of facilitating trade; they work with U.S. exporters—particularly in the agricultural sector—to help them allay the pest concerns of the countries that are importing U.S. commodities. Speaking to a Senate committee in 1997, Dan Glickman, the U.S. Secretary of Agriculture, said, "Exports climbed to $59.8 billion in 1996. . . .Today we are the world's leading exporter of agricultural products, commanding a 23 percent share of world agricultural trade—up from 17 percent a decade ago. Our agricultural trade surplus totaled $27 billion in 1996—the largest in history—making the agricultural sector the largest positive contributor to the U.S. balance of trade." An APHIS study estimates that "almost seven billion dollars in U.S. agricultural exports were protected from potential . . . barriers by the intervention of APHIS and other Agencies." APHIS officials argue that supporting exports doesn't detract from their prevention duties, that they're merely leveraging their expertise and infrastructure.

The flip side of certifying exports is certifying imports. This approach to prevention remains minor compared with port inspections, but it's growing, and both APHIS and its critics embrace the idea. Schwalbe points to the case of Dutch bulbs. For more than 50 years the Dutch, under the scrutiny of APHIS, have made sure that every shipment of bulbs headed for U.S. ports meets acceptable standards. Once certified as low-risk, these shipments pass through U.S. ports subject only to spot-checking, better preventing invasion and facilitating trade at the same time.

Eplee likes this new trend. He points out that it's difficult to deal with shipments after they've arrived at a U.S. port: expenses start piling up, transporters want to get moving, rejecting a shipment causes a furor. It's far easier to inspect or stop a shipment at the point of origin. "Preclearance," says Eplee, "is just a more logical and effective and probably far, far cheaper way to deal with things than for us to try to be a little policeman out there snooping through bags and containers." Preclearance also gives the U.S. the opportunity to shift the expense of inspection from American taxpayers to the traders. Eplee even sees the potential for APHIS, should their agriculture and bug biases ever moderate, to use the preclearance permit system to require exporters to screen shipments for invasive plants and pests of natural areas.

Any substantial expansion of APHIS into new arenas seems unlikely right now. The agency is desperately trying to cope with the burden imposed by booming international trade—imports grew 52 percent from 1990 to 1995—and their charge to facilitate that boom. The 1997 GAO report discovered that APHIS inspectors have been forced into taking many shortcuts. Schwalbe talked about inspectors examining as little as 2 percent of a shipment. GAO found that at one busy border crossing with Mexico, inspectors only managed to inspect about one-tenth of one percent of the passenger vehicles. At another border crossing, inspectors aren't even present for 10 of the 18 hours during which the crossing is open each day. At 52 "low-risk" border crossings, no APHIS personnel ever are present; immigration and customs officials are left to try to do the inspections for exotic invaders.

GAO inspectors encountered the influence of the free traders, too. Large amounts of cut flowers come into Miami, the second busiest port in the U.S. APHIS considers cut flowers a high-risk cargo. Yet according to the GAO report, "The APHIS port director said that inspectors are able to conduct only cursory inspections during high-volume periods because flowers are perishable and the cut-flower industry has continually pressured both political representatives and APHIS to have inspections performed more quickly." GAO officials even "observed inspectors allowing import brokers of cut flowers to select samples for inspection."

Efforts to prevent the entry of invasive species into the U.S. and to quickly eradicate some of the pests that slip through the net are one aspect of the control of invasive species. But what happens when a nasty exotic becomes firmly established on U.S. soil and can't be eradicated? Who stops it from spreading from state to state?

Sometimes the answer is APHIS. Other times the answer is no one. If the species in question is an insect or pathogen that harms agriculture, and is not spreading so rapidly that it's probably beyond control, APHIS likely will act. They typically impose a domestic quarantine that restricts the interstate transport of the species or of items that might carry the pest. If the invader is a plant or a species that primarily affects natural areas, no one is likely to act, at least not anyone in a position to effectively keep the invader from spreading.

People who fight weeds are particularly prone to teeth-gnashing when discussing APHIS's refusal to restrict the interstate movement of invasive weeds. That refusal even extends to weeds on APHIS's noxious-weed list. This creates bewildering situations in which APHIS tries to keep a prohibited species out of the country, but will not prohibit that species from moving within the country. APHIS, for example, will stop a truck attempting to carry a noxious weed from Mexico across the border into Texas, but it won't stop trucks carrying that same weed from Virginia to Texas. A study done in 1990 revealed that at least nine plants on the federal noxious-weed list were being sold in interstate commerce. "We [APHIS] think that once we get a species into the United States, we're already infested," said Eplee. "Well, what percentage of the area is infested? One thousandth of one half of one percent? So, big deal. Does that mean that we've got to capitulate the rest of the country to that species simply because it got a foothold and we don't want to do something about it?"

APHIS administrators say they can regulate interstate transport of an invader only if they first impose a quarantine, and they can impose a quarantine only if a control program exists and the invader is not yet widespread. Since APHIS began administering the act in 1975, they've imposed only one domestic quarantine, and that was on witchweed, an invader that could have devastated the U.S. corn industry. Critics from all quarters disagree with APHIS's interpretation of the Noxious Weed Act and contend that it gives APHIS enough authority to regulate interstate transport, though everyone agrees that the act should be revised to clarify the agency's mandate. Eplee says that one of the main purposes of the Noxious Weed Act was to give APHIS authority over the interstate movement of invasive plants.

"There's a nursery up in Maryland that uses hydrilla as filler material when they send out aquatic species all over the country," Eplee explains, "and we [APHIS] have not done one thing to thwart them, except to tell them they ought not to do that." He added, "That's like the EPA saying [to a chemical company] 'You ought not to sell that chemical. It's bad on the environment. Please don't sell it.'"

When I asked Schwalbe about that Maryland nursery spreading hydrilla, which was placed on the noxious-weed list in 1976, he said that as a rule APHIS does not try to regulate species that are already so widely established. I asked Schwalbe what if the nursery ships this weed to places that are 500 miles from the nearest pond choked with hydrilla, thereby infesting virgin territory? Doesn't APHIS see this "widespread" rule as a weakness they'd like to correct? "To some degree, yes," he replied. "But we need to ask if regulations are indeed the best way to prevent a pest from spreading."

He brought up purple loosestrife, an exotic that has taken over wetlands from Maine to Oregon. Conservationists, biologists, and land managers have been clamoring for APHIS to place loosestrife on the noxious-weed list and then stop nurseries from shipping it from state to state. But, asked Schwalbe, would regulating something as widely distributed as loosestrife really do any good? I pointed out that many I had talked to felt that restricting interstate sales would at least slow its spread, maybe delay or even prevent its arrival in the many wetlands it still hasn't corrupted. Scientists in fact appear to have developed a highly effective biocontrol program for loosestrife. The cavalry may be coming.

After a pause, Schwalbe shifted gears back to hydrilla. He pointed out that in order to stop the shipments containing hydrilla, APHIS would have to issue a quarantine for the area. That would force the nursery to apply for a permit to ship hydrilla from the area, a permit which APHIS could then deny—but consider the difficulty of enforcing the quarantine, said Schwalbe, and the waste of scarce resources on this one problem.

I imagine that most of the people in charge at APHIS would dearly love for things to be the way they were 10 or 20 years ago, when APHIS could quietly pursue its chosen mission of protecting certain agricultural sectors from certain exotic insects and diseases. In this regard, APHIS has done and continues to do a decent job. But now, alien battles have turned into a full-scale invasion, and APHIS, with an assist from the U.S. Fish and Wildlife Service, provides our main line of defense. The Forest Service, the BLM, the National Park Service, state and county governments, private land owners—they all can try to control exotics on their lands, but only APHIS and Fish and Wildlife have means to prevent invasive species from entering our nation and from spreading state to state.

Certainly, APHIS faces obstacles: The agency could use more money, clearer authority, and more freedom from trade agreements. But APHIS shouldn't let these obstacles stop it from carrying out its unique mission.

CHAPTER 10

AVANT-GARDEN

E ven if APHIS was capable of preventing the entry of all pro-hibited exotics and of all those that are unlisted but unwanted, the invasion would continue. By no means are all exotic species unwanted. Many individuals, businesses, and government agencies spend a great deal of time and money procuring alien organisms. They bring in tarantulas, geckos, black mambas, and hissing cockroaches as pets. They stock Sika deer from Asia and South African oryx on private ranches so hunters can go on safari without a passport. They use exotic grasses and shrubs to stabilize soils and feed livestock. They populate their aquariums—and sometimes our waters—with neon tetra, piranha, ripsaw catfish, and giant snakehead. They bring in Japanese oysters, African fish, and Asian shrimp for the aquaculture industry. They import African clawed frogs for medical experiments. The list of intentional introductions is long and getting longer.

Over the years, gardeners and other plant aficionados have been among the most prolific importers of exotic species. According to a Gallup poll, about 70 million American households engage in gardening. For some that may amount to nothing more than a monthly ramble on a riding mower, but for many people gardening means ornamental horticulture. These latter gardeners seek plants that are beautiful, fragrant, striking, or unusual. Often they want something new, and something new usually means something that's not native, and something non-native sometimes means something invasive.

Melaleuca. Purple loosestrife. Scotch broom. Australian pine. Kudzu. Andean pampas grass. Brazilian pepper. Water hyacinth. English ivy. Multiflora rose. Shoebutton ardisia. Tamarisk. Cogon grass. European gorse. Chinese tallow. These are some of the most destructive invasive plants in the United States. Most were brought to this country as garden ornamentals, all have been distributed as ornamentals, and most still are being distributed as ornamentals. The ornamental trade has been and continues to be a major pathway for the entry and establishment of weeds. There are virtually no restrictions on what species can be acquired. In *Invasive Plants: Weeds of the Global Garden,* a 1996 book published by the Brooklyn Botanic Garden in New York, the authors calculate that of the 300 most invasive exotic plants, about half have been brought here to decorate the nation's gardens. Peter S. White, director of the North Carolina Botanical Garden, at the University of North Carolina, Chapel Hill, estimates that 85 percent of all the invasive weeds in North America have been imported for horticulture or other landscape purposes.

"One of the botanical gardens' historic missions," said White, "has been to explore the world for useful plants and bring them back and display them in the garden. And, if they work out well, to encourage nurseries to grow them and sell them to the public. So you have an effective distribution center for plants, some of which become invasive in natural landscapes." Collecting and distributing exotic plants remains a defining mission for most botanical gardens. And they've been notoriously slow to come to grips with the alien invasion.

As soon as I parked the car, I spotted the first invasive species, a bamboo, growing outside the garden's fence and clearly free to escape into the surrounding woods. I'd come to tour Kanapaha Botanical Gardens, a 62-acre facility on the outskirts of Gainesville operated by a nonprofit, the North Florida Botanical Society. Bamboos are a specialty at Kanapaha; it boasts the largest public collection in Florida. The garden collects bamboo species from all around the world, fourteen of which were available at the annual winter plant sale. A garden brochure explains that bamboos are the world's fastest growing vascular plants, able to reach speeds of more than an inch an hour. Meant to wow visitors, that little-known fact sends chills down the spines of people who fight invasive plants.

As I walked across the parking lot, I saw some cogon grass, a major-league invader that overruns pastures, pine plantations, and a variety of natural areas throughout the Gulf Coast states. Cogon grass has been ranked as one of the world's ten worst weeds and has made it onto both the federal and one of the Florida noxious-weed lists. This knee-high grass from southeast Asia has spread to Kanapaha from the surrounding wilds.

At the entrance I encountered catclaw mimosa. The little sign informing visitors that this mimosa hailed from the West Indies and Argentina indicated an imperative that drives most botanic gardens: the tendency to gather things from far away. Catclaw mimosa is one of 62 plants that made it into category one on the 1997 list of Florida's most invasive weeds. This list was compiled by the Florida Exotic Pest Plant Council (EPPC), an influential group of scientists and land managers who work on invasive plant issues. EPPC divides its list into category one—"Species that are invading and disrupting native plant communities in Florida"; and category two—"Species that have shown a potential to disrupt native plant communities."

Fifteen feet inside the entrance, there was an invasive privet, a handsome shrub often used as a hedge. Three species of privet plague Florida's wildlands; one species made EPPC's category one, the others have been placed in category two. Just beyond the privet I saw lantana, whose little flowers are prized by butterflies and by people who prize butterflies. Lantana is a category-one invader. Presently I came across a passiflora species known as stinking passion-flower. A category-two plant, this one is an invader in Hawaii, which makes it worth keeping a close eye on in the sunshine state. Likewise, I felt a little uneasy about a ginger I spotted, because some gingers have run wild in Hawaii. Not far from the ginger I passed a category-one species, Japanese honeysuckle—a major invader, particularly in the Northeast, the South, and the Midwest.

Kanapaha, like almost all botanical gardens, is loaded with invasive exotics. Plants move beyond the confines of botanical gardens in many ways. Most botanical gardens sell seeds and spores through the mail, though usually to each other. Peter White reports that many gardens are proud of the fact that they ship seeds and spores all over the nation and the world. Some species escape without help from human hands. Birds, rodents, water, or the wind may carry seeds and spores from botanical gardens to nearby natural areas. Urban botanical gardens may feel exempt from these concerns, except the mail-order pathway, because they're far from any natural areas. John Randall, the Nature Conservancy's invasive-weeds specialist and a leading authority on plant invasions, finds this argument unconvincing, even if the garden lies in the middle of New York City. "The person in Brooklyn gives [the plant] to his daughter-in-law in New Jersey, and she gives it to her brother in Pennsylvania, and boom, it's out there."

Most important, botanical gardens do much of the foreign exploration. Nurseries and gardeners often look to the botanical gardens for new species, and for leadership and scientific authority. By helping fuel the alien invasion, botanical gardens give legitimacy to the dangerous status quo in the world of plants, in which pretty much anything goes—and comes.

That most botanical gardens are oblivious to the issue of invasive species surprised me. I expected them to be vividly aware of the dangers, leaders of the charge to repel the invaders. I knew that many botanical gardens had embraced a conservation ethic and were instrumental in protecting native plants, especially rare and endangered species. And I knew that invasive plants constitute one of the main threats to native plants. I, therefore, assumed botanical gardens would be committed foes of the alien invasion.

The reasons they are not may have to do with not fully appreciating the significance of the problem and with a reluctance to discontinue the sales of exotic plants that have been big moneymakers. But I also think that confronting the full meaning of the exotic invasion threatens the culture of botanical gardens. Roaming the world to collect new plants, bringing them home to display, giving them to the horticultural trade and to the public—for decades that has been what botanical gardens do. Facing their complicity in the alien invasion also challenges their sense of identity. Botanic gardeners see themselves as "green." They supply living beauty and wonder. They forge deep ties to the Earth. It's difficult for them to accept that some of their actions cause great harm to the natural world they cherish.

Recently Fairchild Tropical Gardens near Miami brought in several world-renowned authorities to help them plot their course for the 21st century. The esteemed consultants identified conservation as the appropriate focus for the next 100 years. Botanical gardeners see themselves as charter members of the conservation community. Propagating endangered plants, for example, comes easy to most botanical gardens, a natural complement to their historic mission. But, added the authorities, the number one conservation issue on which Fairchild should focus is invasive plants.

Not surprisingly, Fairchild didn't immediately embrace the consultants' report. "That's like taking them from kindergarten to graduate school in one week," said Carol Lippincott. A botanist who worked at Fairchild in the early 1990s, Lippincott introduced something alien to the garden: the idea that invasive plants are a serious problem. She admits that others who worried about the invasion may have come on too strong. "You can't shove this down people's throats. There are really hard decisions that have to be made, and botanic gardens have to work their way gradually toward them."

Contemplating the prospect of a major transition for botanical gardens, Lippincott said, "You know, I'm a hopeful person. You have to be if you're in this field [invasive plants], or you'd just crawl into a corner and drink yourself to death." She paused to laugh, then added, "I continue to hope that what I'm seeing are growing pains, and that in a decade, maybe two, botanic gardens will be the leaders in the crusade." Perhaps Lippincott won't have to wait that long. Recently, having had time to digest the consultants' report, Fairchild has moved vigorously to change their ways. They've stopped selling any plant species suspected of being invasive, and they're planning to hire someone to develop an anti-invasives program.

Nurseries link consumers to botanical gardens and to the other institutions that collect and develop plants. They provide the main conduit by which invasive ornamentals spread. Once nurseries have sold a invasive species to thousands of home gardeners and landscapers, it is bound to escape into the wild. And nurseries are big business and getting bigger.

"The average person across the country doesn't realize the magnitude of the nursery industry," said Earl Wells, executive vice president of the Florida Nurserymen and Growers Association. In 1996, nursery and greenhouse crops ranked seventh in total grower cash receipts among all farm commodities in the nation, according to a U.S. Department of Agriculture study. Rising steadily for the last two decades, those cash receipts amounted to nearly $10 billion in 1996—about 10 percent of the total for all U.S. farm crops.

The USDA figures reveal some of the quantitative; Hallie Dozier's research provides a glimpse of the qualitative. Originally trained as a social scientist, Dozier is now completing a doctorate at the School of Forest Resources and Conservation at the University of Florida. In 1996, Dozier conducted a survey of nursery customers in Florida, Louisiana, and Texas, exploring their buying habits and their knowledge of and attitudes toward invasive exotic plants. She found that six out of seven plants bought were exotic and that about 75 percent of nursery stock was non-native. These numbers no doubt differ in different parts of the country, but the general point applies to nurseries across the nation: An overwhelming majority of the plants they stock and sell are exotic.

Why do American nurseries mostly deal in plants from other continents? Partly out of the tradition of overseas plant collecting and partly from a more practical conservatism. Nursery owners and employees invest much time and money in breeding a particular plant species, in learning how to propagate it efficiently, and in learning how to handle it. Naturally they're reluctant to stop selling that plant, even if it turns out to be invasive. Nurseries also cultivate their customers, many of whom return for favored exotics time after time.

No one knows just how many of the exotics that nurseries carry are invasive. In some cases, only time will tell. In other cases, the research hasn't been done. Even when scientific evidence does exist and it damns a particular species, nursery owners and employees probably aren't aware of it. At each nursery Dozier visited, she would locate the most knowledgeable person at the nursery, whether owner or employee, and ask about the origins of the plants being sold. On average, for six out of ten plants, this person couldn't even name the continent or general region from which the species came. Their lack of awareness probably has an enormous ripple effect. According to Dozier, studies show that the most important source of horticultural information for a gardener is his nursery.

The nursery industry can take some comfort in the fact that a vast majority of exotic plants are not seriously invasive; there's apparently no need to worry about selling exotic azaleas or camellias. Yet even a small minority amounts to quite a few species when you consider how many thousands of non-native plants nurseries sell. And it's likely that the percentage of invasives stocked by nurseries is higher than the percentage of invasives among plants in general. "The plants that are invasive that are ornamentals often are popular because they're so easy to grow," said Dozier. "People come into a nursery and say, 'Give me something that's green, that flowers, that's pretty, and I can't kill it.' The same qualities that make plants desirable ornamentals can also make them weedy invasives. They're brought in for those very reasons." Gardeners should bear in mind, however, that while an invasive plant may establish easily and be impossible to kill, it may require lots of maintenance because it grows so prolifically. Think of the constant trimming that English ivy requires.

For years there has been conflict between the nursery industry and people worried about plant invasions, Florida being a good example. The Florida nursery business—second only to California's—is worth more than $1 billion a year. Florida also is the second most invaded state in the union after Hawaii, and some of its most troublesome invaders have been sold by nurseries. And continue to be. About 35 percent of the species on EPPC's 1997 category-one list are available at Florida nurseries.

The powerful nursery industry has mostly shrugged off attempts to address the issue of invasive plants. But sometimes pesky David has riled Goliath into action. A couple of years ago, a nurseryman in Florida set out to legally circumvent restrictions on selling water hyacinth. Florida had long ago prohibited the propagation and sale within the state of this terrible invader, and the feds had forbidden its sale to other states. But this nurseryman hoped to sell it to dealers in Canada who in turn would sell it to customers in the U.S. The rationale for this enterprise was that sales would be limited to northern climes, where the cold would supposedly prevent the sun-loving hyacinth from escaping cultivation.

The nurseryman got on the phone to the Florida statehouse, and shortly thereafter, legislation was proposed that would clear the way for such a sale. The Florida Department of Environmental Protection (DEP) discouraged this risky business; they felt that hyacinth might adapt and become invasive up north. They also feared that it might find its way south. Convinced that the risks far outweighed the potential gain, the DEP tried to dissuade the legislature from furthering the nurseryman's plan. Not only did the DEP fail, but in the process the nursery industry and their boosters in the legislature nearly brought down the Bureau of Aquatic Plant Management, a branch of the DEP.

The hyacinth case represents an extreme event, but the friction it revealed has been all too prevalent. Many in the horticulture industry wouldn't mind seeing the bureau and other hotbeds of anti-invasive activity disappear. As the most vocal opponent of invasive exotics, EPPC has borne the brunt of the criticism. Some in the industry portray EPPC as a fringe group whose members think all exotic plants are bad and should be replaced by natives.

Though some EPPC members have occasionally taken extreme positions, most of the several hundred EPPC members are government land managers, agency scientists, or university professors and researchers. They work at Everglades National Park, the Southwest Florida Water Management District, Sarasota County, the University of Florida, and the state parks. Several past board members came from chemical companies, including DowElanco. And EPPC is well aware that most exotics are not bad.

The industry makes a number of reasonable objections about EPPC's approach. For example, Earl Wells notes that EPPC's list is too inflexible. He points to nandina, a highly profitable nursery shrub that also is known as heavenly bamboo, though it's not a bamboo. Nandina is a category-one invader on EPPC's list, and Wells grants that it can be invasive, but he says it's only a problem way up in northern Florida, not in central and south. Nancy Coile agrees with Wells, both about the need for regional distinctions and about nandina being a headache only in northern Florida. "We've got nandina growing outside our door here, and it's not a problem." "Here" is at the Gainesville office of the Division of Plant Industry, where she works as a botanist. Hearing that Coile works for this unit of the Florida Department of Agriculture, a staunch defender of the nursery industry, people might expect her to agree with Wells. On the other hand, people might be surprised to learn that Coile belongs to EPPC.

Coile also agrees with Wells when he says that EPPC shouldn't condemn all lantanas, another big seller for nurseries, when some varieties are sterile and not invasive. She also agrees with EPPC that invasive lantana is bad news and shouldn't be sold: "I think it's a weed that's really going to be a problem," she said. "There are nurseries that have quite a large stock of [the invasive] lantana, and they'd probably be willing to hang me from the rafters for saying this." She adds, "That's my opinion as an EPPC member, a botanist, and a person who sees the damage being done to the environment by invasive plants." Though she agrees with a few of the nursery industry's complaints about the EPPC list, Coile thinks that overall the list is quite accurate and useful. "I like having the list," she said. "It calls attention to plants that I think are going to be a real problem or already are a real problem." Glancing down the EPPC list, which she keeps in her office, she said, "Most of the ones on the list aren't commercially or horticulturally important. Just nasty weeds that nobody wants."

The EPPC list does mention nandina, lantana, and a number of other nursery standbys, however, and that has made it a lightning rod for conflict. You may recall that EPPC ranks 62 species as category-one plants. Of those 62, only 14, most of them weeds that cause economic harm, appear on the three official rosters of prohibited species: the federal noxious-weed list and Florida's own two lists. The Florida Department of Environmental Protection has attempted to add some of the worst offenders to the state lists, but the Florida Department of Agriculture has veto power, a power they've repeatedly exercised to protect the nursery industry. Despite years of effort by agencies and individuals concerned about natural areas, only a few species have moved from the EPPC list to the state list.

Chinese tallow is one of the few species that made the prohibited list. Nurseries in Florida and throughout the Southeast like this Asian import for landscaping. It quickly grows to 30 or 40 feet and its leaves turn yellow and red in autumn, providing fall color in a region starved for that autumnal gift. Conservation agencies hate Chinese tallow. It thrives in a wide range of habitats, from suburbs to wet prairies to bottomland forests, in sunlight and in shade, in fresh water and in brackish water. Its seeds disperse readily via birds and water. Its dropped leaves exude toxins that change the soil chemistry. It promotes non-native soil arthropods at the expense of native soil denizens. Tallow has taken over whole bayous and large tracts of coastal prairie in Louisiana and Texas, and altogether it has infested nine states, including Florida. Conservationists have dubbed tallow the "North Florida melaleuca."

Connie Riherd, assistant director of the Division of Plant Industry, touts tallow as an example of the nursery industry acting as a good environmental citizen. She says that in the mid-1990s, "the nursery industry recognized that [tallow] was a problem. We met with Earl Wells, his board members, and said 'This plant shouldn't be propagated in the state. It shouldn't be sold through the nursery industry.'" Alan Shapiro, the immediate past president of the Florida Nurserymen and Growers Association, also offered tallow to me as an example of the industry doing the right thing. Shapiro owns San Felasco Nurseries, a large operation located in Gainesville. "Most nurserymen realize that if a plant is a pest plant, we have no business growing it." He added, "When they came to us about Chinese tallow we told them we have no problem phasing it out. We did ask that they not say that it's illegal to sell it today and you have to eat your 2,000 plants that you have, because we can sell it in other states where it isn't a problem." Indeed the phase-out the growers wanted took place. Starting January 1, 1996, they had two years to sell out their inventory before the full-fledged prohibition on tallow kicked in.

Worries about tallow surfaced long before the mid-1990s, however. People from the DEP grew concerned over Chinese tallow way back in the early 1980s. By the mid-1980s, they were talking about the need to deal with tallow. They held workshops, published warning pamphlets, and generally put out the word. DEP repeatedly informed Agriculture Department officials about the dangers of tallow and asked that tallow be added to the state's lists of prohibited plants. The agriculture department refused. Tallow was a good plant for nurseries; it didn't invade farmers' fields—what was the problem? Only a couple of years ago did the agriculture department relent.

Not only did the industry's good citizenship take a long time to develop, but they took too full advantage of the phase-out period. Instead of backing off tallow as soon as possible, they produced it by the truckload and aggressively marketed it. Many tallow trees were sold in South Florida, where little tallow had previously been planted.

The industry's altruism in listing tallow was laced with pragmatism. According to Wells, during the tallow fracas, the DEP was also talking about putting lantana on the prohibited list. So the industry did some horse trading and finally struck a deal to give up tallow, a relatively minor product, if they could keep lantana, a bestseller.

The situation in Florida had the makings of prolonged trench warfare, but suddenly in late 1997, a truce was signed. Wells approached EPPC with an invitation to talk. Wells, Shapiro, Greg Jubinsky, and a couple of others subsequently met. The talk went smoothly and they've continued meeting in 1998. Jubinsky reports that the talks have been respectful and productive. "We just wanted to open the lines of communication," said Jubinsky. "Both Earl [Wells] and I thought that we need to take baby steps at first." For starters, they'll be looking for win-win species. Wells has sent out a survey to some of his organization's members to help identify species on the EPPC list that nurseries consider commercially viable, which in turn helps identify those species that wouldn't be missed much if nurseries stopped selling them. Wells and Jubinsky both think that most of the EPPC category-one species are not big moneymakers, which is promising. Both the industry and EPPC representatives realize that ironing out some of their differences could take years and that some may never be resolved, but calling a truce to work on a lasting peace constitutes a heartening milestone. "We decided to dance together," said Wells, chuckling, "but we haven't decided to go home together yet."

I've heard several theories as to the reasons behind this unexpected rapprochement. Some say EPPC has softened its rhetoric significantly, realizing they'll never get much of anything listed if they keep trying to get too many things listed. Others say the industry fears regulation, and prefers to work out an accommodation more to their liking than what the government is likely to impose. Probably there's truth to both these speculations. But many other observers pointed to a more fundamental factor. Though the tug-of-war over invasive plants appears to be a good, old-fashioned struggle between industry and the environment, there's a big difference: the nurseries are not the equivalent of oil companies or strip-mining operations. Like the staff at botanical gardens, most nursery owners and workers love plants. They like to grow things, feel the soil in their hands. Most of them are green at heart, and as they learn more about the alien invasion, they'll be motivated to help prevent further harmful introductions. Certainly they won't be willing to sacrifice their businesses, but they won't have to. There's a lot of fertile middle ground in this debate.

People have begun exploring that middle ground at the national level, most notably a small but influential group that first met in 1997. Representatives from the nursery industry, the conservation community, APHIS, the scientific community, botanical gardens, and state governments got together to talk about invasive plants and the role of the plant industry. This landmark meeting indicates that the issue is starting to get the attention it requires. Much of the impetus for forming this working group came from the American Nursery and Landscape Association (ANLA), the nation's main trade organization for this industry, and its director of regulatory affairs, Craig Regelbrugge, who co-chairs the group. The ANLA's interest in invasives echoes the Florida nursery industry: they want to protect their business from overly restrictive regulations and they have come to realize that invasive plants constitute a major environmental problem that nurseries and landscapers can help address.

"When I first joined the staff here in 1989, this issue was not on the radar screen," said Regelbrugge. As the first blips appeared, they were indeed perceived as incoming enemy attackers. Regelbrugge said a split exists between the horticulture community and the environmental community, and they haven't been exchanging ambassadors. "There hasn't been a lot of communication. And some of the messengers have been people whom [nursery and landscapers] haven't wanted to hear the message from—people who have a very strong view that the nursery industry has single-handedly mucked up the ecology on this continent."

Industry members likewise have reacted defensively to attempts to make the use of natives mandatory, such as an effort to institute a "natives only" policy for all federal and federally funded projects. "That's a quick way to get plant people to bristle," said Regelbrugge.

"The industry's message is that plant selection ought to be based on a number of criteria, especially analysis of the site and the desired end result." In a nutshell, "Right plant, right place." If it's a rural highway, says Regelbrugge, try to use native plants. But if it's a city street, it may be that only an exotic will work, though preferably one that's not invasive. "We need to pay attention to invasiveness . . . but it's only one factor."

As communication between the industry and the anti-invasives crowd has improved, progress has been made. "I think more and more people [in the trade] are taking the issue seriously," said Regelbrugge. He cites a Massachusetts nurseryman who dropped some invasive species from his catalogue; a Minnesota grower who agreed to destroy his crop of buckthorn after he learned that it was a nasty invader of natural areas; and a project being conducted by the Horticultural Research Institute to assess voluntary methods to reduce distribution of invasive plants. Regelbrugge hopes that these scattered examples represent the beginning of a concerted effort to add "invasive" to the work-a-day vocabulary of the nursery and landscape professions. "Invasiveness needs to work itself into the decision-making process as a criterion, just as one ought to be considering pest complexes that'll affect a particular plant, sun or shade, and ornamental attributes. Invasiveness ought to be every bit as prominent in the plant-selection process."

Making consideration of invasiveness part of the selection process is the immediate goal of the working group Regelbrugge co-chairs. The key is to develop a screening process that will predict whether a plant someone wants to bring into the U.S. will be invasive. Pure researchers in the field of exotics must raise their eyebrows at the working group's lofty ambition. Predicting invasiveness has been something of a holy grail, and an elegant theory explaining the nature of invasiveness remains elusive. But Sarah Reichard wasn't searching for the grand theory of invasion biology when she developed her process. A research assistant professor at the University of Washington's Center for Urban Horticulture, Reichard set out to devise something more modest and yet immensely important: a relatively simple, practical method for making a reasonably accurate assessment of a plant's potential for invasiveness. She finished her model and, after modifications, presented the idea to the working group. By the end of 1998, they hope to start promoting her protocol to nurseries and landscapers, asking for their voluntary participation.

Scientists have identified four general ways to deal with introductions. One is the current method: people import any plant they want unless it's one of the few on the federal noxious-weed list. This reactive strategy typically bans a plant after it has proven itself to be a horrible invader—which usually means it's too late to get rid of it. Another strategy would be to ban all exotic plants. This is unworkable, unnecessary, and would propel nursery operators into the streets brandishing their hedge clippers. A third possibility would be to experimentally test all proposed import species for invasiveness, but that would cost huge sums of money and take lots of time. As Regelbrugge said, "You'd have to do 30 years of grow-out trials to get results." The last way, Reichard's way, is to learn from the past.

Reichard studied hundreds of exotic woody-plant species that had entered the U.S., including some that had been in the country for at least 60 years without incident, some that had escaped cultivation and become established in the wild but apparently without doing any harm, and some that were harming native species and ecosystems. She examined a variety of traits in each of these plants, such as the length of the flowering period, the mechanism for dispersing seeds, and the germination requirements. Key characteristics began to emerge like clues in a crime. Fifty-three percent of the harmful exotic species spread both vegetatively and by sexual reproduction, but only 23 percent of the noninvaders did. The seeds of 70 percent of the noninvaders require some kind of pretreatment (such as being chilled by cold weather) before they'll germinate, whereas only 49 percent of the invaders must clear this hurdle.

Reichard also looked beyond the plants' biological traits. She studied some geography and discovered that 25 percent of the noninvaders are fellow North Americans, but only 3 percent of the invasive species came from another part of North America. She also found out which plants are invasive elsewhere in the world. This common-sense line of inquiry turned out to be one of the most valuable. Reichard learned that only 15 percent of the noninvaders are invasive somewhere else, while 61 percent of the harmful species invade elsewhere.

No single factor provides a high degree of predictiveness, but Reichard combined a number of the most telling traits into a decision-making model. Thus a species that reproduced both vegetatively and sexually, that produced seeds that didn't need any pretreatment, and that was invasive elsewhere would rank as a very likely invader. Her model led to three possible outcomes: acceptance of the species in question (it showed a low probability of being invasive), rejection (a high probability of being invasive), and hold for monitoring (indications of potential invasiveness). To test her model, she compared the well-known woody species against her screen. It worked great for picking out the invaders. The screen rejected 85 percent of the invasive woody plants, it recommended that 13 percent be held for monitoring, and it mistakenly gave the green light to only two percent of the invaders, none of which was one of the harmful invasive species. When she tested the noninvaders, the screen accepted 46 percent of them, rejected 18 percent, and recommended holding 36 percent for monitoring. These results aren't as accurate as are those for the invasive species, but that means the model would err on the side of caution in terms of protecting the environment, which seems prudent, while still making available plenty of new exotic plant species.

Reichard is the first to agree that her process is hardly perfect; the complex realm of invasion biology defies perfect understanding. The imperfection also goes beyond the scientific obstacles; inevitably, values and assumptions color decision-making models. For example, Reichard injected some subjectivity into the validation of her model when she labeled certain invasive woody species as "harmful." Science can't be expected to provide all the answers to questions that are largely unscientific. In fact, Reichard unabashedly adds personal recommendations along with her scientific work. For example, she suggests that if screening becomes mandatory, importers should not be held liable if a species that passed the protocol turns out to be invasive later. And she emphasizes that "the economic value of the species *must* be secondary to the assessment of invasive potential."

No doubt Reichard and others will refine her screening process as it gets used. Other models exist or are being developed for other classes of organisms. For instance, more than 25 years ago the American Fisheries Society produced a protocol on new fish introductions, though virtually no agencies or individuals have seen fit to use it. It's healthy to have a variety of models, because in this endeavor one size does not fit all. Now that reliable and practical models are emerging, the next big step will be putting them to use. As always, voluntary compliance serves everyone best; but if people won't do what needs doing, we'll eventually have to consider regulation. These protocols won't prevent the entry of the next melaleuca if they're sitting on a shelf gathering dust.

Botanical gardens, nurseries, landscaping firms, and the various government agencies that bring in exotics all play vital roles in preventing the intentional introduction of invasive plants. But consumers form the base of the pyramid, the foundation upon which rests the structure that discovers, develops, and imports exotic plants. We form the base of other pyramids, too. We eagerly catch the rainbow trout that the U.S. Fish and Wildlife Service sows in Lake Mohave. We press the California Department of Fish and Game to provide more wild pigs for us to hunt. We take home—and sometimes improperly release— the exotic snakes, fish, insects, and other critters rounded up by the pet industry.

While we were unwittingly responsible for the introduction of many invasive species, we now have the power to change the way the future unfolds.

Why do people buy the plants they buy? Hallie Dozier, the Florida researcher who conducted the survey of nurseries and nursery customers, phrased the answer this way: "color, color, and color." Her data shows that other factors mattered, notably ease of care, but, above all, buyers wanted something colorful. This might suggest something bright and tropical, something exotic, but that's all right. There are plenty of bright, tropical plants that are not invasive. Nowhere did customers say that the plant also had to be invasive or they wouldn't buy it. Other surveys in other parts of the U.S. might have discovered other traits that customers look for in their plant purchases, but no doubt these traits, too, could be found in noninvasive exotic species.

Native plants also can provide most of those desirable qualities but in most cases haven't been given a fair chance to satisfy the needs of gardeners and landscapers. In many cases we think of certain exotic plants as irreplaceable simply because we haven't tried to replace them. Chinese tallow, for example, is prized by many southerners in large part for its fall colors. A brochure prepared by the Florida Department of Environmental Protection and EPPC lists 44 native trees that provide fall color.

Those native species, however, might be lacking in other qualities that nurseries and customers desire. Many of the plants distributed by botanical gardens and sold by nurseries have been altered, bred for easy propagation, for showier blooms, or for any number of other attributes. This is the essence of cultivation, and it often gives long-time favorites, most of which are exotics, an advantage. But it's not an inherent advantage. "If the ornamental industry had put as much emphasis on improving native plants," said Dozier, "so that they would be better horticultural selections, maybe we would have really good native substitutes. They have been researching and working on geraniums and begonias for at least a hundred years, selecting superior plants or hybridizing them. Horticulturalists haven't been doing that with most native plants." She paused to think of an example. "I don't understand why they're not working on making a horticulturally acceptable variety of American beauty berry, which is a gorgeous wild shrub around here. But it's real leggy, and it doesn't make a nice, compact shrub." Those are qualities, she added, that plant breeders probably could change. The industry would do well to pay attention to Dozier's survey, which found that nursery customers wanted more natives from which to choose. "I asked them if they'd like to see more native plants in nurseries," said Dozier, "and they overwhelmingly said yes."

Craig Regelbrugge, the American Nursery and Landscape Association executive, agrees that providing substitutes is important to the nursery and landscaping industries: "If there are other plants that essentially fill the same niche in landscape design, then that eases the pain [of giving up an invasive but profitable species]." The idea of cultivating alternatives is not limited to natives. The essential point is to find noninvasive substitutes, whether native or exotic, for invasive plants. "I think we'll make more progress in the public-policy arena," said Regelbrugge, "if we can take a step away from the native-origin question and focus more on the invasives question."

If gardeners are to motivate nurseries to develop and stock more noninvasive plants, first the gardeners themselves must be motivated. Awareness of the larger importance of the alien invasion is the key. Most gardeners still think of invasiveness in personal terms and don't realize that invasive plants do enormous environmental damage. Dozier touched on this angle in her survey: "I asked if they'd like to see nurseries label plants as to weediness, and they said absolutely yes." But, she added, "I don't think people are thinking beyond their own backyards." However, Dozier believes that gardeners care about the environment and would be likely to respond if they knew more. "They're not malicious spreaders of nasty weeds."

Numerous gardening organizations recently have taken on the task of spreading the word about the invasion. "Awareness is the first step," said Patricia Schleuning, national horticulture chairman of the Garden Club of America (GCA). When she tells members about the dangers of invasive plants, "They're shocked. They had no idea." Schleuning understands that coming to grips with the issue can take time, and that gardeners sometimes struggle to part with favorite species. "One of my friends," said Schleuning, "who's a wonderful horticulturalist, takes me through her garden every now and then. Last summer as I went through, she said 'I'm ashamed to show you that my beautiful purple loosestrife is in bloom. I know I should remove it, but I don't have the heart to do it yet because it's so pretty.'" Now there's a situation made to order for a worthy, noninvasive substitute.

In 1998, GCA sent its members a booklet entitled *Choices for Gardeners.* It briefly sounds the warning about the dangers of invasive plants and asks its members to educate themselves. "In today's world," reads the booklet, "the basics of gardening need to include knowledge of invasive non-native plants, particularly the ornamentals that are escaping from our gardens." The booklet suggests ways in which gardeners can help, such as by eliminating known invasives from their gardens and talking to neighbors and friends about the issue. The bulk of the booklet is devoted to a state-by-state listing of hundreds of invasive exotic plants and native alternatives to each of those invaders. Those in the nursery industry who find this talk of native alternatives unsettling will take heart at one of the booklet's messages: "The majority of exotics are well-behaved, deserving a place in our gardens."

Regional, state, and local gardening organizations also have taken on the education of their members. In 1996, a regional committee of GCA published a booklet called *Exotic Ornamentals: Innocent or Invasive?* which goes into considerable detail about both the general aspects of the invasion and specific invasive plants that plague that region. One passage that merits quoting appears under the heading: "Why Should I Be Concerned? My pet plant is well-behaved!" The answer: "Many of us have plants in our gardens that we are shocked to see on invasive lists. It may be our favorite Porcelain Vine growing in a container by the kitchen door, a clump of *Miscanthus* [a grass] by the pool, or a hedge of lovely rugosa roses. 'Why are they targeted?' we ask. 'The Porcelain Vine can't go anywhere in a pot, there's not a blade of Miscanthus anywhere on my property, and the roses are unique on my street! How can these plants be invasive?'

"Well, how about the birds that eat those Porcelain berries? The wind that carries the grass seeds afar? Or the rose hips that fall and are swept into the gutter to be carried into the storm basin? Sometimes, it's just a matter of time for a plant to adapt to local conditions before it 'escapes' the managed cultivation of the garden and is set loose into unmanaged areas. How many of us have seen a plant struggle season after season until one day it 'just takes off'?"

Like the national office, the regional GCA urges its members to talk up the issue among friends and neighbors. In addition, the regional committee's publication asks members to discuss invasiveness with nurseries, departments of transportation, garden clubs, and other buyers and sellers of plants. A sort of gentle activism seems to be welling up among gardeners. In Virginia, I encountered a polite but firm watchdog by the name of Jocelyn Sladen. A member of both GCA and the Virginia Native Plant Society, this petite, white-haired woman keeps a sharp eye out for invasive plants. In 1997, she noticed that her local soil and water conservation district was offering people a free seed mix for growing plants for wildlife that included a serious invasive exotic. She wrote a letter to the district director informing him of the drawbacks of spreading seed from this or any other invasive weed. Sladen also showed me a GCA postcard that people can send to the publishers of seed catalogues. The card leaves a blank space on which to write the name of the offending plant, identifies the plant as invasive in the region, and then states: "I hope you will take this information into account in determining whether or not to continue offering this plant."

John Randall, the invasive-weed specialist for the Nature Conservancy, has encountered an increasing interest among gardeners. "I think awareness is starting, and if we [the anti-invasive advocates] continue to be reasonable and not nasty, I think pretty quickly awareness among people in garden clubs and other organizations like that will grow." Randall thinks that the nursery industry has softened its don't-mess-with-my-exotics posture in part because members of the industry also perceive this trend in the gardening public. In addition to shunning invasives and urging others to do likewise, gardeners can help in another important way, according to Randall. "I'd like to see really advanced gardeners, if they do plant something new, or even something old, keeping an eye on it. If it does start to escape, I'd like them to tell groups like the EPPCs. [Other states and regions have exotic pest plant councils, too, patterned after Florida's.] Then we can get the word out about a new problem in short order. Especially where it's a new species, a novelty, gardeners could be the first line of stopping the problem and become part of the solution."

The spirit of grassroots responsibility rippling through the gardening world is cropping up elsewhere as people try to prevent introductions. In Butte, Montana, car washes and sports clubs wash the undersides of hunters' and fishers' vehicles to keep them from spreading range weeds. To limit the spread of zebra mussels, people who go boating in the Mississippi River, which is infested, clean and inspect their watercraft before entering clean waters. The park service asks visitors to some of their weed-prone parks, especially backcountry hikers, to clean out the grooves in their hiking boots before moving from one area to another.

As awareness of the alien invasion grows, perhaps the little acts of prevention will become part of the fabric of everyday life, like not tossing litter out the car window or extinguishing your campfire. In fact, in the 1960s the U.S. Department of Agriculture tried to raise the public's awareness of pest invaders by borrowing a page from the Forest Service's successful anti-fire campaign built around Smoky the Bear. Their symbol was "Pestina." The fact that not one in a million Americans has ever heard of Pestina reveals how effective this effort was, but the basic idea of making invasiveness a household concern remains valid.

CHAPTER 11

CONTROL

Above our heads the sun flashed through the tattered canopy of redwoods, oaks, and maples. Below our feet the cast-off leaves of autumn crackled and crunched. Shirt-sleeve, Indian-summer weather and the luxuriant natural beauty of a Big Sur glen elevated our short walk from outing to idyll.

While heading down the central California coast to a symposium on invasive plants, my three companions and I had stopped off at Andrew Molera State Park to meet with Judith Goodman, a local conservationist. She was taking us to look at the vegetation project on which she and a few other area residents had been working. When we reached the half-acre site, Goodman and I stood in the little clearing and talked about her group's effort to remove French broom from the hillside. My companions listened briefly, but compelled by a zeal whose depths are hard for the uninitiated to appreciate, they soon gravitated to a nearby stand of the handsome, basketball-hoop-high exotic shrubs.

Goodman and I continued to talk, but we grew increasingly distracted. There was Dan Gluesenkamp, hanging from the bowed top of a strapping French broom plant, seeking better leverage as he struggled to rip the two-inch-thick stem out of the ground. Behind him, flailing about in a dense thicket, Greg Gaar braved abundant poison oak in order to savage some offending shrubs. A few yards up the slope, Jake Sigg knelt beside a specimen too big to uproot. With his knife he deftly carved long strips of bark from around its base, mortally girdling it.

Gripped by the prevailing fervor, Goodman joined in. Wielding a Weed Wrench™ the petite Goodman began prying out broom with a vengeance.

Broom-bashing represents control at its crudest: you see an invader, you destroy it. The art and science of controlling invasive exotics ranges from the brutish level of the broom-basher to the most sophisticated reaches of molecular biology. We need that broad range of approaches, given the hundreds of different invasive species that we must deal with. We also know more invaders are coming. No matter how well APHIS does its job or how conscientious most gardeners are about abstaining from the use of beautiful invasives, some aliens will make it here—and then we'll have to do something about those newcomers, too.

Bashing non-native weeds is becoming a standard item in the community-improvement repertoire, like picking up litter and planting trees. In my town in Oregon, there's a church that occasionally organizes its members to yank Scotch broom from a nearby park. Our public-works department sometimes forms citizen work parties to remove Himalayan blackberry from the banks of a creek that runs through town. I've encountered similar efforts all over the nation. In several towns in Florida, volunteer groups calling themselves "pepper busters" go out on most weekends to bust Brazilian pepper. In Hawaii, the Nature Conservancy guides citizens into the depths of the rain forest, where they uproot kahili ginger and other unwelcome intruders. On Roosevelt Island, located in the middle of the Potomac River in the shadow of our nation's capital, people armed with clippers and trowels have waded into thickets of English ivy, joined by political luminaries, such as Bruce Babbitt, the Secretary of the Interior.

Unfortunately even when fortified by the presence of such august personages, weed bashers face severe limitations. There's only so much that a small group of people can accomplish by hand, even if their pulling-power is mechanically multiplied by Weed Wrenches™ or chain saws. Near San Francisco, in the Golden Gate National

Recreation Area, volunteers logged more than 20,000 hours over the course of several years. Yet they were able to clear and keep clear only about 60 acres. Using muscle power and hand tools to tackle vast exotic blight, such as leafy spurge creates, would be like trying to clean up a toxic-waste dump with a toothbrush.

But weed-bashing has its uses, such as removing invaders from circumscribed areas that harbor endangered native plants. These surgical strikes are especially helpful for cases in which early eradication is still possible. As any worthy weed warrior will say as often as someone will listen, we've got to expunge invasive plants when they first show up, before the population machine gets cranking. If we did the necessary surveying to spot budding infestations when they still consisted of but a few plants, hand-pulling could be effective. Getting after weeds on an up-close-and-personal level serves another important function. It builds awareness of the alien invasion, gives volunteers a visceral feeling for the nature of the problem—and can be deeply satisfying, too. I once labored with a team in the Golden Gate National Recreation Area. One middle-aged guy told me that he relished doing something more concrete than writing a letter to a senator. A young woman, busy levering out a sizeable French broom plant, said that feeling the roots tear and finally give way produces an almost euphoric pleasure. And I must admit that I thoroughly enjoyed the weird chemistry of weed-bashing, a blend of community service, camaraderie, invigorating outdoor work, and mayhem.

Physical control occasionally extends beyond the powers of human beings and a few hand tools. Mowing machines and prescribed fire can also be used in certain circumstances to knock back invasive weeds, which significantly increases the size of the area that can be cleansed. If you've got plenty of money, you even can take on some pretty hefty infestations. The biggest physical control project I've ever heard about bears the strange name Hole-in-the-Donut.

The Hole-in-the-Donut is a 10,000-acre block in the middle of Everglades National Park that was farmed for 35 years until the park acquired the land in the 1970s. About 5,000 acres of the block were rock-plowed (the limestone bedrock was broken up) in order to produce a deeper, richer soil suitable for crops. When the farming ceased, Brazilian pepper rapidly invaded those 5,000 plowed acres, forming dense stands in many places.

People brought this shrubby South American tree into Florida in the nineteenth century for ornamental purposes. Also known as Florida holly and Christmas berry, Brazilian pepper was particularly prized by gardeners for its deep green leaves and red berries. Nurseries distributed Brazilian pepper widely until its invasive qualities became all too obvious and the state placed it on the prohibited list. But pepper has continued spreading on its own; though slowed in some places by aggressive management, it now infests roughly 800,000 acres from north of Lake Okeechobee to Everglades National Park. It especially likes artificial habitat such as roadsides, canal banks, and abandoned agricultural fields.

Robert Doren, the assistant research director for Everglades National Park and the originator of the project to restore the Hole-in-the-Donut, took me out to the site on a sweltering summer afternoon. We passed acre after acre of Brazilian pepper thickets, impenetrable tangles of these multiple-stemmed monsters rising to heights of 15 to 25 feet. This infestation could swallow 100 teams of pepper busters without a trace. But Doren and company were taking mechanical control to new heights—and depths.

After a few minutes Doren and I came to a large opening that was free of Brazilian pepper: the first stages of the restoration project. After years of trying all sorts of treatments with little success, an unexpected pulse of funds enabled Doren and company to try something different. "There's nothing you can do until you change one of the fundamental ecological functions," said Doren, "So that's what we did." Talk about getting down to basics. They brought in bulldozers and scraped a couple of hundred acres right down to bedrock. "We went in and we changed state variables," said Doren. "We said, 'No more soil moisture to get [pepper] through the dry season. No more soil to provide the structure to grow.' And by removing the soil, we reduced the nutrients and added higher water. We changed four state variables. People come in and say, 'This is so drastic! Bulldozers and trucks!' Well, think what it took to turn this site into [a Brazilian pepper infestation]. It was even more drastic. Tractors and farming and fertilizers and pesticides . . . for 20 to 30 years."

If the project managers had simply cleared off the Brazilian pepper, without changing the fundamental qualities that made the site so inviting, pepper would have recolonized the treated area. But now the soil structure that allowed pepper to prosper is gone. And scraping off the soil lowered the site's elevation by six to eight inches, which causes the seasonal floods to linger longer than pepper likes. Monitoring of the bulldozed area shows that no Brazilian pepper seedlings are surviving. Zero.

We stepped out into an area that had been scraped clean just two months earlier, in late May 1997. "When you walked out here at the end of May," said Doren, "you walked on solid limestone rock. It was as white as it could be." Now a slimy, brown-green mat of algae, bacteria, fungi, and other tiny organisms squished beneath our boots: new soil being born. Without the rock plowing and fertilizers, this soil would never become the artificially thick and rich topsoil that supported crops—and Brazilian pepper. Instead it would become the thin, rather poor soil natural to this kind of site in this part of the Everglades. Already some plants have taken root in the protosoil. If the land where Doren and I were standing follows the path of an experimental plot that was bulldozed in 1989, it will develop into a short hydroperiod marsh and prairie dominated by wetland grasses and sedges—much the same habitat that occupied Hole-in-the-Donut before farming made it the hole-in-the-Everglades. If this restoration project continues as planned, all of the Brazilian pepper should be gone from the Hole-in-the-Donut by 2016—and the native ecosystem should be well on its way to recovery.

The Hole-in-the-Donut restoration certainly looks as if it will turn out to be a success story, but before resource managers rush to try this at home, they had best take note of the price tag: $60 million, in 1992 dollars. Of course, the costs from one situation to the next aren't so easily compared; but the general point holds: manual or mechanical control is prohibitively expensive in most cases.

Which plunges us squarely into a frothing controversy.

In those cases when strong backs and bulldozers aren't feasible, the ready alternative is chemical control—all those laboratory concoctions that kill things dead. Insecticides, herbicides, fungicides, and other chemicals grouped together under the term "pesticide": crotoxyphos, methoxychlor, hexachlorobenzene, 2,4-dichlorophenoxyacetic acid. These and other tongue-twisting ingredients go into tens of thousands of pesticides, which appear on the market with friendly names like Round-up® and Sonar.

Research has made it clear that some of these chemicals harm humans and the environment. How they cause harm and how much harm they cause often are controversial questions whose answers, to the extent that there are answers, lie beyond the scope of this book. But concern over the side effects of chemical pesticides pervades discussions about controlling invasive species.

Recent events at Lake Davis illustrate a typical conflict. In recent years, northern pike had taken over this reservoir in the Sierra Nevada in northern California. Not only did these large, aggressive, non-native predators gulp down most of the fish in the lake, but the pike likely would have expanded downstream to other parts of the Sacramento River system. Conservationists and state officials worried in particular about the impact pike would have had on the beleaguered native salmon populations. So the state decided to eradicate the lake's pike. As usual, the managers turned to chemicals, in this case Nusyn-Noxfish and rotenone. And, as usual, a number of people didn't want to see 33,000 gallons of chemicals pumped into a lake.

Local residents led the outcry. Lake Davis supplies much of their drinking water, and they worry that hazardous chemicals may be coming out of their kitchen faucets. They're especially concerned about the Nusyn-Noxfish, which contains tiny amounts of trichloroethylene, an organic solvent that may cause cancer under some conditions. Officials and a number of scientists assured the locals that there was no danger. The state would supply drinking water from an alternate source until health officials determined that the chemicals had broken down and the lake water once again was safe to drink.

These assurances didn't assure some local residents. They took their opposition to court. The court eventually ruled in favor of the California Department of Fish and Game, and the agency began preparations to poison the lake on October 15, 1997. On September 16, the locals tried another angle, adopting a county ordinance making it a misdemeanor to pollute a body of water by intentionally killing fish; officials expected the chemical treatment to result in about 70 tons of dead fish. This ploy, however, also failed to deter Fish and Game. Locals escalated to more aggressive tactics, threatening to barricade roads and to form human chains around the lake. In the end, four locals, including a city councilman, chained themselves to a buoy in the lake early in the morning on October 15. Later that day, the protesters were removed and the chemicals were put in. Conservationists who had worried about the salmon were happy; conservationists who had worried about the chemicals were unhappy; conservationists who had worried about both felt conflicted. Fish and Game officials and some members of the local tourist industry who wanted to stock the lake with rainbow trout, were happy; many the locals, who felt victimized and unsafe, were unhappy. And the pike and all the other fish in Lake Davis were dead.

Only the manufacturers and sellers of Nusyn-Noxfish and powdered rotenone came out ahead. Such unsatisfactory scenarios play out over and over across the nation as farmers, government agencies, homeowners, ranchers, timber companies, and many others use about two billion pounds of chemical pesticides a year. And that's two billion pounds of active ingredients. Some modern chemicals are so potent that just a tablespoon of the active ingredient will treat a whole acre.

Not all of those billions of pounds of chemical pesticides are applied only to controlling invasive exotics. But many are. Virtually everyone who battles alien invaders considers the use of chemicals to be a lesser evil than allowing exotics to run wild. Even the scientists and land managers trying to protect natural areas, who typically express more concern about pesticides than do people in agriculture, consider the use of chemicals a grim necessity. Most of these natural-area managers sympathize with people who object to pesticides, but they think that the protests should be tempered by a better understanding of the environmental dangers posed by an uncontrolled alien invasion. Don Schmitz, an aquatic biologist with the Florida Department of Environmental Protection, asked a tough question, "What the hell do they really want? Do they want a short-term environmental insult or a long-term ecological catastrophe?"

Mary O'Brien rejects the idea that one must either rely on chemicals or lose the battle with alien invaders. A botanist and former university professor, O'Brien for many years has been a leading antipesticide crusader. She has served as a consultant to the United Nations, the Department of Defense, and various regional and state organizations. Many times in her career she has encountered that either-or idea. It's a controversy that she often faces in her current work in Hells Canyon, where invasive range weeds are wreaking havoc. "The issue gets framed as, 'Use herbicides or let [invasive weeds] spread,'" said O'Brien. She sees pesticides as only one of many options—preferably the last one. Her research has led her to believe that heavy chemical use can inflict serious damage on the very ecosystem that the chemicals are supposed to be saving. "I'd have to say that there's so much evidence of both direct and indirect harm to ecological systems that it's very clear that [using pesticides] is like hitting a nail with an anvil.

"Too often the spraying is done as a sign that we're doing something, never mind that it's going to be a pesticide treadmill. You spray. It leaves bare ground. You continue the practices that favor weeds, such as livestock grazing in native bunchgrasses, which didn't evolve in the presence of large numbers of large grazers. The weeds come back. You spray. You leave bare ground. How far-thinking is that?" O'Brien thinks we need to control invasive species not only by dealing with the invaders but by dealing with the ecosystem being invaded. "You examine the conditions that favor the introduction and spread of noxious weeds," she said, referring to the example of Hells Canyon, "and you don't favor them. You manage land differently, so that you don't continually make conditions perfect for noxious weeds. That's your first line of defense." She added, "This is what a lot of land managers resist, because a lot of prevention is doing things differently than they do. They don't want to rock the boat in terms of livestock grazing, clearcutting, whatever. They don't want to change that. Then they want to hand the public . . . the choice of either letting noxious weeds spread or spraying them.

"I think the single largest barrier to managing with less herbicides . . . is that you have to know the land better. That you have to know the ecological system. The use of pesticides allows you to go at an ecosystem about which you know very little. You read the label, you get the recommendation from the extension service . . . and you go at it." O'Brien notes that organic farmers often learn from books written in the 1920s and 1930s, before the chemical revolution erased the knowledge of how to control invasives without using tons of pesticides, but, she admits, "this takes time. And it takes walking out on the land. And it takes mimicking nature rather than subduing it."

Despite her concerns about the use of chemicals, O'Brien acknowledges a role for them, especially in the short term. "I would not say that we could handle the noxious-weed problem we've got now without any pesticides." But she wants pesticide use to be minimal and part of a long-term plan to get off the treadmill. "I'll accept some spraying if it's in the context of a whole, planned program," she said. She thinks that managers sometimes need to use pesticides to get a grip on runaway invasives, but they need to be planning how they'll prevent the situation from recurring. "At the very best," said O'Brien, "[the judicious use of chemicals] can give you a moment, a window of opportunity in which to change your ways."

But fewer and fewer chemicals that could be applied judiciously are being produced. Bernd Blossey, a biological-control specialist from Germany who has been working out of Cornell University since 1992, explains that the expense of developing and registering pesticides steers companies away from producing so-called "minor-use" products. Most chemicals intended primarily for use on environmental invaders or on pests of relatively minor crops fall into this category. The industry mostly caters to "the world's ten big crops," said Blossey. "People talk about the big companies like Monsanto developing something that targets one problem plant in a minor crop. Forget it. No way." The financial incentives dictate the development of broad-spectrum pesticides, each of which kills many species. But these indiscriminate chemicals cause problems in agricultural systems over the long haul and can produce harmful side effects in natural systems from start to finish. This trend dampens the hope of devising enough highly selective chemicals, each of which would target a specific invader without harming desirable species.

Fortunately we can avoid the choice between excessive use of chemical pesticides and the degradation of our ecosystems and cultivated lands by invasive species. Crop rotation, breeding pest-resistant species, and plowing under crop residue, as the farmers in the Imperial Valley have done to deny a safe haven to the whitefly, are alternatives to pesticides that date back many years. Much has been learned by returning to the knowledge possessed by farmers and land managers prior to the ascendance of chemical pesticides in the 1940s. New technologies, lumped together under the heading of biologically-based technologies, or BBTs, also hold great promise.

In 1995, the Congressional Office of Technology Assessment published a report entitled *Biologically Based Technologies for Pest Control.* Among the BBTs the report explores are microbial pesticides, pest behavior-modifying chemicals, and the genetic manipulation of pest populations. Microbial pesticides consist of bacteria, viruses, and other microbes that affect targeted pests; one such formulation, Bt, made from the bacterium *Bacillus thuringiensis,* already enjoys widespread use. Behavior-modifying chemicals disrupt the signals that pass between members of certain pest species; pheromone disrupters, for example, confuse the mating cues and therefore the reproduction of many insects. In the cases of pink bollworm and oriental fruit moth, pheromone applications have proven at least as effective as insecticides. Genetic manipulation involves the release of genetically altered individuals in order to weaken the targeted pest population; for example, mass releases of sterile male screwworms have interfered with the reproduction of this major pest species, essentially eliminating it from the United States.

The patriarch of BBTs is biological control. It's the oldest, most established, most promising, most contentious, and probably the most dangerous of all these technologies. Three types of biocontrol exist, but classical biocontrol is by far the most important. More than 100 years ago, people in the U.S. began importing exotic species and releasing them on farms in an effort to control pests. These early efforts produced some spectacular successes, and biocontrol gained increasing attention during the early 1900s. However, the chemical pesticide boom of the forties shoved biocontrol onto a back burner, where it remained until the last decade or two, when the drawbacks of widespread chemical use have become apparent. Natural-area managers and conservationists in particular find biocontrol attractive. Most of them would like to stop leaning so heavily on chemicals.

Biocontrol could be the keystone of a sustainable, nonpolluting, environmentally benign, and in the long run, relatively inexpensive approach to controlling many invasive species. An appealingly simple logic forms biocontrol's basis. To a large degree an exotic runs wild after it moves to a new home because it leaves behind the predators, parasites, and pathogens that previously held it in check. So it makes sense to recruit some of those predators, parasites, and pathogens and reunite them with that exotic species in the U.S.

Biocontrol is not as easy as it sounds, however, as the case of the European gypsy moth demonstrates. Gypsy moths consume about four million acres of American forest a year; researchers have been looking for a biocontrol agent for this dreaded pest since the late 19th century. They've introduced more than 50 biocontrol species, but none has been successful—though, ironically, scientists have recently rediscovered an alien fungus that was introduced as a biocontrol agent in 1910 which may prove effective in controlling European gypsy moths. The terrible batting average of gypsy-moth biocontrol is extreme, but a large majority of all biocontrol agents have failed to do the job, and many don't even establish self-perpetuating populations. A 1988 study of Hawaii's prolific efforts shows that only 243 out of 679 introduced biocontrol organisms even got established.

Most disturbing, the study found that 86 of those biocontrol agents in Hawaii have attacked species besides their targets. The rosy wolfsnail provides a textbook case of biocontrol gone awry. In 1955, this native of Latin America and the southeastern United States was introduced to Hawaii to prey on the African tree snail, an exotic agricultural pest. This fearsome predator unfortunately prefers native Hawaiian snails, which occur in remarkable diversity; nearly 800 species of nonmarine snails have evolved in these islands. Sometimes gobbling shell and all, the rosy wolfsnail has contributed hugely to the extinction of most of the 15 to 20 endemic species of *Achatinella* snails, pushing this whole genus onto the federal endangered-species list. The wolfsnail has hammered many other Hawaiian snail species as well.

Critics of biocontrol management are not surprised by the turn the rosy wolfsnail took. In 1993, Marc Miller, an associate professor of law at Emory University, and Greg Aplet, a forest ecologist with the Wilderness Society, wrote a widely discussed article about the shortcomings of biocontrol. "In order for predators to be effective," they noted, "they must be aggressive, voracious, and widespread. Diseases must be virulent, and parasites must be deadly. Phytophages (plant eaters) must breed to levels capable of killing entire plants or halting reproduction. All biocontrol organisms must be fecund and competent searchers. These same characteristics are possessed by the worst pests." Miller and Aplet believe that biocontrol agents are especially dangerous because they are permanent; once unleashed, they become part of the ecosystem.

The rosy wolfsnail was brought in to attack an agricultural pest, but it harmed natural-area species. Miller and Aplet point out that biocontrol scientists have focused mainly on crops and other economically valuable species when assessing the safety of biocontrol introductions. Researchers made sure, for example, that wolfsnails wouldn't munch on sugarcane, but their potential for harming native plants and animals wasn't evaluated. Miller and Aplet also report a growing interest in the introduction of exotic biocontrol agents in order to decimate native "pest" species, such as spruce budworms or tent caterpillars, that harm crops or commercially valuable timber. This ill-considered notion fails to appreciate the fundamental role such native species play in ecosystems. Attempts to suppress natural outbreaks of native insects are misguided. They suffer from the same commodity-oriented tunnel vision that led to America's long and ecologically destructive efforts to suppress all wild fires. To kill off native insects by bringing in aggressive exotic species piles folly atop foolishness.

This agricultural bias comes as no surprise. Most of the nation's biocontrol program is tied to the U.S. Department of Agriculture. "They're just too agriculture oriented," said Bernd Blossey, the biocontrol expert at Cornell. "Even though they pay lip service . . . to dealing with environmental issues, if you look at it on the ground with who gets the money to do what, and what's actually happening in the field, forget it. I haven't seen it happening." Blossey is a leading authority on the biological control of purple loosestrife, the invader that degrades wetlands from Massachusetts to Oregon. When Blossey requested the assistance of a USDA scientist whose expertise in genetics would have been a big help, the researcher's superiors "wouldn't let him spend any time on it. They didn't permit anyone to work on loosestrife anymore. It's not an agricultural problem." Blossey notes, however, that in the last couple of years, the department's biocontrol program has started coming out of its agricultural shell.

This bias toward agriculture and the accompanying economic

pressures that push the release of untested or poorly tested agents exacerbate the concern about biocontrol within the conservation community, concern that could hold back biocontrol's development. The OTA report on BBTs cites the USDA effort to use biocontrol to deal with the Russian wheat aphid. In the early 1990s, USDA researchers and managers rapidly introduced twenty-four biocontrol species comprising more than a hundred geographic strains without adequate knowledge of the potential ecological effects. For that matter, the USDA didn't even know enough about how well their hastily chosen agents would carry out their appointed task. Only four of the imported parasites are thought to have become established in wheat fields, and so far they haven't done much to control the aphids.

Blossey shakes his head at the carelessness of the Russian wheat aphid snafu. "They basically went out with a sweep net in Europe and Eurasia and brought back [many] different parasites and predators . . . without going through sufficient safety procedures. They were releasing them into the field without knowing how they would interact with each other and the ecosystem." Blossey notes that the USDA biocontrol program seems to need regulating, but the USDA also happens to be the agency vested with the authority to regulate biocontrol. Blossey thinks that oversight of biocontrol needs to be taken from the U.S. Agriculture Department and placed elsewhere, perhaps with a cross-agency body. "We're not just protecting agribusiness," said Blossey. "We're protecting the national environment. We don't want nontarget effects."

The purple loosestrife program provides a model of a well-conceived, meticulous biocontrol endeavor. In large part, the program was started in the early 1980s by a nationwide group of scientists and land managers, with leadership from people at Cornell, worried about the growing degradation of wetlands by loosestrife. Members of the group had already tried chemicals, mowing, burning, and anything else they could think of, but they weren't able to stop loosestrife. Finally they turned to biocontrol, enlisting the European nonprofit biocontrol outfit for which Blossey was working at the time. From 1986 to 1992, Blossey searched for insects that successfully attacked loosestrife in its European homeland. But mere ferocity toward loosestrife wasn't enough to make the grade. Blossey focused on making the selections ecologically sound, on understanding as precisely as possible the ways in which particular insects affected loosestrife and the plant community in general.

That ecological approach led naturally to the core concern of conservationists and botanists: Would the anointed insects stray from purple loosestrife and cause any significant harm to native plants and to the environment in general? In biocontrol jargon, the "host-specificity" of the candidate insects had to be determined—that is, would they stick to the specific host, in this case purple loosestrife? All responsible biocontrol programs devote much thought and experimentation to host specificity, and a number of rigorous protocols have been created. According to Blossey and other researchers, biocontrol agents of invasive weeds that have passed through such strict screens have never strayed beyond their intended victims to any significant degree. Some critics of biocontrol say otherwise. The centerpiece of their case has been a weevil recruited in the 1960s to control musk thistle that subsequently expanded its diet to include some native thistles. Blossy and several other biocontrol experts point out, however, that the weevil case doesn't apply because the screening tests done in the '60s had predicted that the weevil would attack native thistles. The agriculture-oriented managers simply didn't care. Their concern was controlling the musk thistle that infested farmers' fields.

Numerous factors determine host specificity. The ability of a plant species to support the reproduction of a biocontrol agent is one key. In making his selections, Blossey looked closely at about a hundred insect species, then looked very closely at seven or eight species, and finally chose four. Females of the four species he chose will not lay eggs on plants on which their larvae can't develop, and their larvae only develop on purple loosestrife. They can't digest just any old loosestrife, and certainly not some unrelated plant. So when a foraging female can't find its one and only food, it will die from starvation. During its desperate search it may try other species, but not often. Through taste buds on their feet or via chemical cues from plants, these insects can tell if the plant species on which they land is edible. In the case of Blossey's quartet of insects, if it's not purple loosestrife, it's not edible.

Blossey cautions that testing for reproductive potential alone is not sufficient. This view is seconded by Peter McEvoy, an entomology professor at Oregon State University and a leading authority on host specificity. McEvoy allows that testing has come a long way since the 1960s, when researchers simply presented prospective biocontrol insects with various crop plants to see if the insects would eat them. But he recommends taking the process several steps further in the direction of ecological sophistication. Host-specificity procedures need to examine the influences of climate, habitat, and the genetic makeup of the agent and the host plant. McEvoy also thinks that host-specificity testing, no matter how sophisticated, is not enough. "Although such host specificity tests are necessary . . . ," writes McEvoy, "they are not sufficient because a control organism may harm a non-target organism in a multitude of ways." A biocontrol agent, for example, may do no damage itself, but perhaps it can carry a parasite that would harm a native species. Ecosystems are dauntingly complex, and releasing biocontrol agents into them might cause indirect and long-term consequences. Fortunately most biocontrol programs have become much more cautious, unlike the period from 1890 to 1985, when managers blithely pumped those 679 agents into Hawaii.

Strangely, there have been few studies to determine how well the predictions of the screening protocols stand up in the field. The OTA report notes that many scientists criticized the USDA/Russian wheat-aphid program for neglecting effective follow-up. "Little is known," reads the report, "about the effects, good or bad, of the introduced species on the Russian wheat aphid, on other introduced natural enemies, or on native species and ecosystems." Monitoring usually amounts to little more than a land manager visiting a release site, glancing at the plants to see how they look, and maybe catching some of the biocontrol bugs in a sweep net in order to extrapolate rough population figures. Blossey, McEvoy, and other biocontrol scientists have been calling for much better monitoring. For the purple-loosestrife program, Blossey and his colleagues have developed a standardized process that reveals how purple loosestrife, the ecosystem, and the biocontrol insects are faring. Taking monitoring a step further, Blossey has set up field experiments at release sites, gaining knowledge that will help him refine the purple-loosestrife program as it expands.

In the fall of 1997, I visited Blossey at Cornell, in Ithaca, New York. Stocky, shaggy blond hair to his shoulders, and quick to laugh, Blossey seems at once energetic and relaxed. He rustled up some rubber boots, and we soon were slogging through the shallow water and mud at the edge of a campus pond at which Blossey had released some of his biocontrol insects. He showed me a few sorry-looking loosestrife plants. These hip-high, scraggly specimens bore little resemblance to the hearty, 80-foot-tall behemoths I've seen elsewhere in loosestrife-plagued New York. Crouching beside one of the battered survivors, Blossey showed me the effects of feeding by the two leaf-eating beetles he'd recruited from Europe. They'd chomped almost all the green tissue from the leaves. Scars revealed that the beetles even had scraped the green off the stems, leaving a brown skeleton. Blossey assured me that just a couple of years earlier, the margins of this pond had been choked with head-high loosestrife.

We talked about a hypothesis that Blossey is testing. He says that in Europe the purple loosestrife isn't as tall and robust as it is in the United States. Blossey suspects that purple loosestrife in the U.S. can channel more energy to growth than can its European counterpart. Freed from the pressure of those coevolved native insects, the invading loosestrife can allocate fewer resources to the production of chemicals that defend it against insects and instead devote those resources to growth. In the absence of natural enemies, growth becomes a better evolutionary strategy than wasting energy on unneeded defense mechanisms. For example, a loosestrife plant that grows a little taller and a little faster gets more sunlight, which should help it win the plant-to-plant competition with other purple loosestrife plants.

Such competition constitutes natural selection, survival of the fittest. The offspring of loosestrife plants with genes that promote growth should thrive while the offspring of loosestrife plants with genes that promote chemical defenses should languish. Blossey thinks that 200 years of such selection pressures in the United States have led to populations of purple loosestrife that invest much in growth and little in defense. Though his experiments aren't complete, so far they support his hypothesis. If he's right, the practical implications are considerable. If purple loosestrife in America has shed much of its anti-insect defenses, it might be particularly vulnerable to biocontrol agents.

Blossey believes that the plant-to-plant competition between purple loosestrife and native wetland plants also plays a major role in his biocontrol efforts. Those feeble-looking specimens in the Cornell pond aren't as feeble as they look. Only the annual shoots are dead; in the spring, the healthy root masses will send up new shoots that will prosper and reproduce unless they're chewed on. But if the biocontrol insects can stress the loosestrife, then native plants should be able to outcompete it. They should grow taller and shade out the loosestrife, and take food, water, and other resources away from it. This example shows the importance of understanding not only the biocontrol agents and the target species but the ecosystem in which they interact.

In 1992, Blossey came to the U.S. along with the insects. From its Cornell headquarters, the loosestrife program reached outward, sending bugs to more than 1,000 sites in more than 30 other states. The insects successfully established almost everywhere. After a few years, Blossey began receiving glowing reports that the insects had massacred sizeable tracts of purple loosestrife at some of the early release sites and that native wetland plants were recovering. If carried out vigorously, this biocontrol program appears destined to help enormously in the protection and restoration of the nation's wetlands.

Even as Blossey discussed this success, he noted that biocontrol has limitations. Biocontrol programs take considerable time and money to develop. Biocontrol won't work for all invasive species. The science is still young, and biocontrol practitioners need to proceed cautiously. Finally, as with all control methods, biocontrol works best when incorporated with other tools in an Integrated Pest Management program.

Neither biocontrol nor pesticides nor IPM provides much help in controlling one group of invaders: the vertebrates. Invasive plants and pests cause most of the harm and get most of the attention when it comes to control, but vertebrates like the destructive wild pigs can't be overlooked. Yet few remedies exist. Managers have used contraceptives to help control wild horses and a few other vertebrate species. The park service employed helicopters to net and remove some mountain goats from Olympic National Park. For the most part, however, controlling vertebrates remains a basic business, involving rifles, traps, and poison.

Still, there's room for creativity. In the case of the nutria, for instance, rifles have been replaced by jambalaya recipes.

In the 1930s, E. A. McIlhenny received 13 nutrias from Argentina, their homeland. A nutria looks something like a small beaver, except for its long, ratlike tail and its four emphatically orange incisors. Around 1940, several of McIlhenny's nutrias escaped into the coastal marshes of Louisiana. Nutrias love warm coastal marshes, and they multiplied with abandon. Greg Linscombe, a biologist with the Louisiana Department of Wildlife and Fisheries, says that by the late 1950s these invasive rodents were causing severe damage to rice and sugarcane fields. Federal and state efforts to control nutria failed, but when a market for nutria fur developed in Europe in the early 1960s, complaints from rice and sugarcane growers virtually ceased. For about two decades, trappers took approximately 1.5 million nutrias out of Louisiana coastal marshes every year and the growers had no significant nutria problems.

In the 1980s, however, the market for nutria fur plunged and the Louisiana nutria catch plunged with it to a couple of hundred thousand animals per year. Shortly thereafter, Wildlife and Fisheries biologists learned that nutria used those orange incisors for more than snacking on sugarcane. Reports of devastated marshes began coming in. Linscombe says these herbivores, which damage more vegetation than they consume, tend to linger in one area until they reduce it to bare mud. The technical term for such war zones is an "eat-out." Linscombe has seen eat-outs as large as 1,500 acres. A 1996 survey estimated that nutria had damaged 80,000 acres of coastal marsh. Worst of all, many of Louisiana's coastal marshes are built on highly organic soil, held together only by the roots of the vegetation. When the nutria kill the plants, the soil loosens and the tides wash it away. The marshes can't recover because they no longer exist. "Every time an acre of grass and water converts to just an acre of water, the overall productivity of the system drops," said Linscombe. Everything from important commercial fish species to migrant birds suffers.

Eradicating the millions of nutria that live in those marshes is impossible. Trappers could bring nutria under control, but the fur market remains soft. Buoyed by a federal grant from a fund intended to protect coastal wetlands, Wildlife and Fisheries is trying to develop a meat market for nutria. They figure that if both the meat and the fur have value, trappers will once again keep a lid on the nutria population.

Linscombe reports that nutria are leaner and higher in protein than chicken, turkey, and beef. Their taste is said to lie between rabbit and the dark meat of a turkey. In Europe, nutria appear on the menus of fine restaurants as a delicacy known as *ragondin*. There, nutria are raised in hutches, so when selling to Europeans, Louisiana nutria purveyors will style their product as free-range nutria. Nutria boosters also hope to enter the Asian market. Philippe Parola, a famed Louisiana chef who has cooked for three American presidents, toured Japan in March of 1998, demonstrating the potential of imaginatively prepared nutria.

In marshes far to the northeast of Louisiana, another side of the control issue has come to light in flamboyant fashion.

Whether controlling vertebrates or bugs or weeds, managers often forget to address one vital question: What if I succeed? What if I reduce cheatgrass to a trace element? Will skeletonweed and medusahead, even nastier invasive species, take its place? Eliminating exotics without plans for restoration can be a recipe for further degradation.

Restoration must go beyond merely replanting or restocking with natives or noninvasive exotics, though that is important. Ecological processes also must be revived. Fire regimes, flood cycles, and their kin need to be returned to as natural a state as possible. The whole control effort will be wasted if the system remains profoundly disturbed and once again falls prey to invaders.

In those northeastern marshes, both a long-absent native species and an integral natural process are being restored by a single program. Had you canoed through the coves and salt marshes of the Honga River in the winter of 1997-1998, deep in the Maryland backwaters of the eastern shore of Chesapeake Bay, you might have heard an unfamiliar sound. This loud, resonant honking was familiar to people who canoed here 200 years ago, but since the early 1800s, the clarion call of the trumpeter swan has been missing. Hunters, commercial feather collectors, and the loss of wetlands annihilated trumpeters in the eastern United States, including perhaps 100,000 in Chesapeake Bay. But a few days before Christmas 1997, three trumpeters brought that soul-stirring sound back to the Atlantic flyway. Thousands of their brethren may one day follow, once again adding their bugles to the music of the marshes.

The presence of the trumpeters in the Chesapeake today, though noteworthy, is not as significant as the way in which they got there. They flew. At first blush that hardly seems remarkable for a bird with an eight-foot wingspan, but this trio of the continent's largest native waterfowl flew nearly a hundred miles, from Airlie Center, near Warrenton, Virginia. And if they flew back to Airlie in the spring, as was expected, then those flights constitute a migration, which has been the crucial process missing from attempts to restore these native swans.

Trumpeters learn to migrate. Their parents teach them the routes, the navigational skills, and the knack of locating good stopover sites.

After an absence of nearly two centuries, the knowledge of migration along the Atlantic flyway was long lost. If conservationists had merely transplanted a few trumpeters from the flocks that still exist in the western U.S., the swans, untutored and adrift, would have just stayed year-round on the Honga.

Foregoing their migration would not be a trivial matter. Trumpeters belong in the ecosystems of the Chesapeake and other marshy areas back east. But as the swans and the habitat evolved together, migration was part of the relationship. Eliminate migration and the swans' relationship with the habitat would change. For example, a full population of trumpeters that didn't migrate would soon overgraze the submerged aquatic vegetation that supports them and other organisms. Such habitat degradation has in fact happened in the Chesapeake marshes, except the culprits are mute swans, an invasive species from Europe that doesn't migrate.

So how does one teach trumpeter swans to migrate? Bill Sladen asked himself that question a long time ago, and in 1990 he found an answer. In his mid-70s now, Sladen continues to expand upon his uncommonly full life as a medical doctor; a renowned expert on swans, penguins, and geese; a polar explorer; and a card-carrying eccentric. When he learned about Bill Lishman's efforts to teach geese to migrate by following his ultralight plane, Sladen stumbled on his answer. Lishman is the Canadian sculptor whose exploits formed the basis for the popular 1996 movie *Fly Away Home*. In 1990, Lishman and Sladen talked about imprinting trumpeters on the ultralight and teaching them to migrate, but complications forced them to switch to Canada geese. In 1993, when Lishman finally made the epic journey depicted in *Fly Away Home,* he led his "ultrageese" to Airlie, Sladen's domain in the wooded foothills of the Blue Ridge Mountains. Some of the web-footed stars of the movie still grace the pond behind Sladen's house.

In 1994 and 1995, several more flocks of ultrageese journeyed behind ultralights piloted by Lishman or Airlie personnel. Most important, some of the geese began migrating on their own. This success rekindled Sladen's desire to use the ultralight technique with trumpeters. When he finally received the necessary permits, he teamed up with the national conservation group Defenders of Wildlife to form the Migratory Bird Project. Their first mission: restore migrating trumpeters to the Atlantic flyway.

Two months before the three trumpeter swans made their historic flight to the Honga River, I watched them and other swans training at Airlie. At that point, seven swans between four and five months old were candidates to become ultraswans. They had already imprinted upon biologist and pilot Gavin Shire, who was flying the ultralight that evening. The aircraft essentially is a hang glider with a flimsy body suspended beneath the sail and a tiny engine that powers the big fan mounted to the rear. Shire started the ultralight by jerking on its pull cord and headed to the end of the little runway. Presently the ultra-swan hopefuls arrived—in a trailer. Mostly gray, with some white mottling, these youths already had attained adult stature, though they hadn't yet put on all of the 30 or 35 pounds of a mature trumpeter. Neither had they developed their trumpets yet; the squawks and bleats of these youngsters bore little resemblance to their elders' fanfare. The runt of the flock, Primo, hadn't even started flying yet. "I hope he's not afraid of heights," said a young boy who had accompanied one of the Defenders biologists. The swans' caretakers hauled them down to the end of the runway and let them out of the trailer. Moments later, the ultralight whined past us as it gained altitude and circled back to collect the trumpeters.

As the aircraft came in behind the swans at about 30 miles per hour, Shire shouted "Hey, hey, come on" several times; apparently trumpeters respond to their parents' calls, and shouting was Shire's version of honking. One of the huge birds labored into the air and followed the ultralight for about five seconds, then veered off and returned to the others. Shire came around again. This time two swans took to the air and fell into formation just off the left wing of the plane. He called to them on and off, and they stayed with him for about a minute as he curved slowly back to the rest of the flock. After seeing so many invasive species and the harm they cause, I relished watching those hopeful young trumpeter swans cruising on powerful wings through the crisp autumn sky.

But even sweet thoughts of trumpeters soon lead to sour thoughts of alien invaders. The trumpeters' fate seems likely to get tangled with that of the mute swan. Currently there's enough habitat for both species, but that may change if trumpeters flourish and mutes aren't controlled. Right now mutes are thriving. On the Atlantic flyway alone, the population jumped from about 5,800 in 1986 to more than 10,000 in 1996. Maryland has the worst infestation. Its population of mutes rocketed from about 250 ten years ago to about 3,000 today. And because mutes don't migrate, their impacts are hugely magnified.

Mutes also are highly territorial, lording over realms as large as 25 acres. Fierce creatures, male mutes will defend their territories against almost any intruder: native wildlife, dogs, humans, even boats. They'll certainly go after other mute swans that trespass; mutes are one of the few species that will fight each other to the death. They use their powerful wings as clubs. I heard many accounts of mutes breaking the arms and cutting the skin of biologists, boaters, and waterfront homeowners. More than once, mutes have struck jet skiers and knocked them right off their seats. Twice they've even killed people.

Compared with whiteflies or cheatgrass, mute swans are easy to control. Big, visible, and relatively few in number, they even could be eradicated. But one state wildlife department after another that has attempted to control mute swans has been pilloried by the more strident animal-rights groups. Any report of a swan being killed galvanizes a national network of animal-rights activists that brings intense pressure to bear on stunned wildlife officials. Activists vehemently take their crusade to city councils, state legislatures, and the federal government. Some activists unabashedly play to the media, too, which plays up the demonstrations, accusations, and raw passion typical of the animal-welfare extremists' protest theater. Seldom do control efforts continue, and undoubtedly the chilling effect prevents many control efforts from ever beginning. Even live trapping and relocation programs face passionate resistance from the most intense individuals and groups.

Conservationists and wildlife biologists try to explain the situation. They tell the activists about the perils of alien species. They describe the destruction in the marshes, the harm done to other wildlife, the ecological repercussions. But the objections of most mute swan advocates seem to come from the heart, not the head. People love these majestic birds. While in Maryland's mute-swan territory, I saw planters in the form of swans, swan lawn ornaments, porch steps flanked by statues of swans, and stores that featured swans as their emblems. One of Maryland's most outspoken defenders of mute swans even questions the reintroduction of trumpeters because she finds mutes much more beautiful.

It is no small matter that protests from animal-welfare groups and anti-pesticide advocates hamstring many needed control measures, and keep many environmentalists from tackling the alien invasion. Greg Aplet sees this situation from both sides. As a longtime employee of the Wilderness Society, he understands the environmental community's qualms. As an ecologist who realizes the gravity of the alien invasion, he understands the need for the environmental community to rally its troops to combat the invasion. "The environmental community is starting to pick up on the issue," said Aplet. "But it's not a standard part of our message yet." Aplet attributes some of this lag to a simple lack of awareness; the issue is new compared with pollution, logging, habitat loss, and other traditional concerns. But he acknowledges other reasons. "The steps that might be necessary to do something about [invasive species] are troubling to people. This is something that calls for active management. Active management of anything scares [environmentalists]. We don't have a lot of faith in the agencies to do a very good job of active management. We suspect that they'll twist it toward some commercial end." Aplet thinks that many environmentalists also harbor deeper fears of active management. "The need to do anything at all undermines our faith in nature," said

Aplet. "We've always thought that if we could just protect the land, nature would do the rest." But in the case of invasives, acknowledges Aplet, active management is required.

Environmentalists also know what active management usually means when dealing with invasive plants. "People are locked into an older issue," said Aplet, "a fear of chemical treatment, a fear of herbicides, pesticides. The idea that we might actually need some spot applications of herbicides in wilderness, for instance, in order to stem the invasion, scares the hell out of people." Aplet divides this fear into two categories: fear of the chemicals per se, and fear of the distasteful task of selling the idea of occasional pesticide use to their members.

"Another thing that makes people uneasy about it is consorting with people who just like pesticides," said Aplet. "I went to [a meeting of a weed group] at the office of DowElanco, and there were two guys from Dow and one guy from Monsanto and I think I was the only enviro. . . ." Chuckling at the memory, he continued: "At one point I asked one of the guys from Dow, 'What's your interest here?' And he said, 'Well, we just think that if we can raise awareness of this [issue], we'll sell more herbicides.'" Such comments elicit groans from purists in the environmental community. Recognizing pesticides as a legitimate and important tool is hard for them.

Aplet shifted gears to animal rights. "That's one that people are really afraid to touch, too," he said. But he thinks environmentalists understand this issue better because it's easier to see the problem of a landscape denuded by exotic animals than one invaded by alien plants. "When you see a chain of goats tromping through the dust on Catalina Island, you know that's wrong." At least at the Wilderness Society, said Aplet, "we're less concerned with the rights of individual animals than with the extinction of entire species." But he does encounter individuals in his own organization who disagree. "I asked [a colleague] point blank one day, 'What would you rather see? The life of an individual [pig] extended a couple of years or a species allowed to survive an eternity?' Well, she had to pick the pig."

"When I've gone to these meetings [about invasives]," said Aplet, "environmentalists have been noticeably absent." But he notes that the Nature Conservancy and, to a lesser extent, the Audubon Society are exceptions. He thinks it's because both organizations, particularly the Conservancy, manage wildlands themselves. When they see yellow starthistle or wild pigs destroying their land, they get less squeamish about the occasional use of unsavory methods to control such invaders.

PART III: THE FUTURE

CHAPTER 12

CHOOSING OUR FUTURE, SAVING OUR PAST

I nvasive exotic species have changed the planet. But the onslaught has not yet turned our world into a place that our ancestors would hardly recognize. The world, though altered, may remain a place brimming with natural riches that support and nourish. It's up to us.

Experts speculate that if we continue to approach the alien invasion as we have been, only a few, isolated native plant and animal communities will remain intact. Agriculture could falter; the cost to American taxpayers and businesses could run into the hundreds of billions of dollars. The capacity of biological communities to respond to environmental changes will be diminished, and basic natural processes, like the cycling of nutrients, will be altered—a change that could degrade essential resources like soil and water. In a dark moment, one biologist, dismayed by the spread of cheatgrass, envisioned a wasteland of bare rock and eroded earth populated only by a horde of spiked weeds.

And what if we do step up our efforts? Again, there's much uncertainty, but biologists and other experts have some predictions. Some think we can keep the native communities in our natural areas largely intact, and that industrial agriculture can find a new, more productive path. Some think we can keep our ecosystems functioning smoothly, and that we can reduce most alien species to a background level and hold them there. The consensus is that we have the means to contain the alien invasion. But do we have the will?

Now that we are becoming aware of the extent of the alien invasion, we're beginning to appreciate what measures need to be taken to combat it. A clearer picture of our future—both the dangers and the opportunities—can be brought into focus by looking at our island state. Hawaii is ahead of the continental U.S. in dealing with invasives because it's ahead of the mainland in suffering from invasives: Hawaii is the exotic species capital of the world.

Driving along the Hana Highway on the northeast coast of Maui, the alien invasion seemed far away. Bob Hobdy, a forestry manager with the Hawaii Department of Land and Natural Resources, sat at the wheel of the pickup, easing us through the tight curves of this famously meandering road. The highway clearly deserves its reputation as one of the world's most scenic drives. Camel-hump mountains, cliff-top views of the wild Pacific, slender waterfalls tumbling into clear pools beside the road, and lush vegetation in infinite shades of green highlighted with the reds and whites and yellows of tropical blossoms can all be experienced along its dramatic length.

"Virtually everything you're seeing is not native," said Hobdy. He pointed out some thickets of eight-foot-tall Brazilian pepper, the dreaded invader that has caused so much concern in the Everglades. Hobdy said that pepper spreads stealthily: "It will start in the under-story," said Hobdy, "and you won't see it, and you won't see it, and then *bam!*—it's everywhere. It overtops the natives, and many Hawaiian [native plants] can't handle being overtopped . . . they die quickly." Hobdy added that the pepper also has rolled over pastures, as have several other exotics, creating major headaches for Hawaiian ranchers.

A little farther down the highway, we spotted some large stands of rose apple, a South American tree noted for creating exceptionally deep shade. "It's like night underneath it," said Hobdy. "Nothing can survive under it except its own seedlings." We passed some molasses grass, an invasive that fuels destructive fires. Here and there we saw strangler figs choking trees, though along the Hana Road these invasive vines have nothing left to strangle but exotics. Hobdy sees this as part of a downward spiral. "After an area is already so completely invaded, anything else that comes along and succeeds is probably bigger and meaner," he explained. We drove by entire hillsides covered by a species of bamboo that had been introduced to bind soil. Hobdy said that the bamboo is expanding exponentially. He pointed out a small patch of catsclaw, a thorny vine that has taken over thousands of acres on Kauai, another of the Hawaiian islands. "I've seen a cow killed by that stuff," said Hobdy. "It got caught in the catsclaw, and the more it thrashed, the more the thorns got it, until it was suspended off the ground and died." He expects catsclaw will thrive on Maui, too. During our two hours on the Hana Highway, Hobdy pointed out dozens of exotic plant species, only a small fraction of the approximately 900 exotic plants that have become established in Hawaii. Nearly half of all the plant species that grow outside cultivation in the island state are non-native, and at lower elevations, where the land has been more heavily disturbed, nearly all of the vegetation is non-indigenous.

As Hobdy and I passed an African tulip tree, he pointed to some birds flitting about its wide-spreading branches. Mynahs. An exotic. The mynahs and a few other bird species were the only non-native animals I noticed along the highway that day, but that's just because the others are less visible. A thorough search would have turned up all sorts of alien animals. Hawaii has no native land mammals except the bat. Nineteen non-native mammals live in the forests, including highly destructive pigs, goats, mongooses, and axis deer. Five native fish species reside in Hawaii's freshwater streams and ponds; twenty-nine exotics have joined them. The state's 13 reptile species and 4 amphibians are all exotics. About 30 percent—2,500 species—of Hawaii's insects are non-native, and on average another 20 exotic insect species add to that tally every year. At least three dozen alien bird species have joined the mynahs in the islands' treetops. Hawaii's animal kingdom has been transformed perhaps even more thoroughly than has its plant kingdom.

When the first humans arrived in Hawaii about A.D. 400, they found a paradise. It wasn't the sugar-sand beaches or the tropical climate that made Hawaii an Eden with few peers; it was what the islands didn't have. No biting insects. No cockroaches, ants, or scorpions. No poisonous spiders and no poisonous snakes—no snakes at all, in fact. No tigers or crocodiles, no mammals or reptiles of any sort, except for one bat species. No poisonous plants lay in wait for unwitting humans. None of the plants had thorns. No malaria or other deadly tropical diseases. Numerous bird species strolled around on the ground, having lost their ability to fly after evolving in an environment that harbored no predators from which they must flee.

Hawaii's geographical isolation accounted for the benign nature of its native flora and fauna. Twenty-five hundred miles from the nearest continent, Hawaii is the most remote cluster of islands on the planet. Before humans arrived, animals and plants could reach Hawaii only by wind, wing, or water. Many species had no way to reach Hawaii. Those few organisms that did reach Hawaii evolved enthusiastically over the eons, branching into many marvelous native species, nearly all of them endemic to the islands.

This miracle of evolution has left most Hawaiian species vulnerable to invasives. Natives had little need to develop the defenses that mainland plants and animals had evolved in order to survive in a competitive world, one in which species prey on each other as they fight for resources. Hawaii's native ground-nesting and flightless birds, for example, were easy prey for the pigs, dogs, cats, mongooses, rats, and other alien mammals that humans brought with them.

The litany of loss sickens the heart. More than one-third of our nation's endangered species live only in the Hawaiian islands. More than 72 percent of our recorded extinctions have occurred there, including 63 bird species, and 75 percent of the remaining bird species are on the endangered and threatened list. About half of Hawaii's fascinating endemic land snails are extinct, and most of the remaining half are imperiled. Approximately 10 percent of the islands' native plant species are extinct, another 30 percent threatened or endangered. About 30 percent of the insects are in jeopardy.

The continental U.S. is not immune to such pervasive problems. Our natives already have tangled with and lost to cheatgrass, melaleuca, the silverleaf whitefly, carp, and many other invaders, which, by virtue of the advantages invasives enjoy, can be overpowering. And while the lower 48 won't experience as many extinctions, exotics are threatening many natives. Consider California's oak grasslands, in which 99 percent of the grass is exotic. No native grass species has gone extinct, but that doesn't help the populations of oak trees, which are in trouble largely because the alien grasses prevent oak seedlings from surviving.

In mid-April 1995, the traditional Polynesian sailing canoe, *Hokule'a*, and its five sister canoes had set sail for Hawaii from French Polynesia. The crew members were retracing the migration route that some of their Polynesian ancestors had taken many centuries before in similar vessels. But just a day short of completing their 6,000-mile voyage, the canoes stopped. Something terrible had happened. No, not a typhoon or a capsized canoe. Crew members had been bitten by midges.

After radioing to Hawaii for advice and consulting with his crew, the *Hokule'a's* master navigator, Nainoa Thompson, and the captains of the other canoes made the decision to come to about 200 miles from the Big Island of Hawaii. They halted at such a distance to ensure that the wind couldn't possibly blow any stray midges to land; then crew members gathered every bit of organic matter on the boats that could harbor midge larvae and heaved it overboard. Meanwhile a Coast Guard plane flew out from Hawaii and airdropped canisters containing cans of aerosol insecticides. Crew members sprayed their big, double-hulled canoes inside and out.

Finally the *Hokule'a* raised its sails and made for Hilo harbor on the Big Island. Outside the port, the voyagers stopped again. A fumigation team met the boats and tented them. Only then did the *Hokule'a* enter the harbor. After the dockside welcoming ceremonies ended and the crew got off, officials thoroughly fumigated the vessels.

The object of all these precautions was a gnatlike insect almost too tiny to see: a midge called a punkie. These wee beasties had stowed away on the sailing canoes in the Marquesas, where the punkies have tormented people for decades. First they use their scissorlike mouth parts to chew holes in their victim's skin, then they lap up the blood as it seeps out. People all too familiar with the punkie's bite say it produces a burning sensation, as if you'd been singed by a lighted match. Their diminutive size makes punkies especially annoying; they can fly right through most screens and mosquito nets, and they easily get under clothing.

Hawaiians shivered at the thought of a punkie invasion. Their sunny, open-air way of life depends in large part on the fact that Hawaii is blessedly free of biting insects. Hawaii's sun-and-fun tourism industry also relies on bug-free beaches, and tourism is the backbone of the state's economy. If bloodthirsty midges swarmed tourists on Waikiki, Hawaii's economy would unravel. Hawaiians expressed gratitude to the voyagers for recognizing the threat posed by those biting stowaways and for going to such great lengths to make sure no punkies reached Hawaii alive. The voyagers displayed the level of awareness and commitment that it's going to take to fight invasive species.

As Bob Hobdy and I were driving along the Hana Highway, it occurred to me that one of the biggest obstacles to raising the level of awareness about alien species lay before me: the distressed landscape looked like paradise—not a profoundly disturbed ecosystem.

Simply telling people about the perils of the invasion will not necessarily produce comprehension and support. Some people place blind faith in what they perceive as nature. They point to an example such as the Hawaiian hawk, which feasts on the exotic rats. "Look!" they say, "This native bird species adjusted to the influx of aliens, other native species will adjust, too. It'll all work out, so we shouldn't fuss over all this invasion hype." This point of view, however, overlooks the fact that many native species will not adjust as well as the hawk has, if they can adjust at all. The many Hawaiian bird species that have gone extinct or are slipping toward the brink are an apt example. And those rats that make such nice prey for Hawaiian hawks have contributed to the extinction and endangerment of other native Hawaiian species.

Another popular rationale for ignoring the alien invasion is that the spread of exotics is just evolution at work—the fittest species will survive. Hard-pressed to explain the high number of natives that are vulnerable on the islands of Hawaii, this theory becomes particularly seductive. But Hawaii's natives are vulnerable because of the rapid change brought on by modern civilization. Prior to the arrival of humans and their alien baggage, these species had survived for eons. We have accelerated the natural movement of plants and animals around the globe, and nowhere more so than in Hawaii. This hardly is the pace of evolution. To label as unfit a species that has not adjusted to a radically altered environment is just as inaccurate as saying that fish that can't survive toxic spills are weak and their demise is simply part of evolutionary development.

One of the looming transformations that will affect alien species is what the 1993 OTA report calls the "wild card": global climate change—a.k.a. global warming. This phenomenon will to some degree reshape the Earth's climate, a shift resulting from changes in the atmosphere's chemical makeup—caused by, among other factors, the emission of greenhouse gases.

Scientists and land managers concerned about non-indigenous species certainly take climate change seriously. It is predicted that the mean temperature of North America will rise several degrees Fahrenheit over the next century, depending on how vigorously people work to reduce emissions. Such a rise will disrupt U.S. ecosystems, setting the stage for further invasion. That temperature rise will draw many new invaders up from Mexico and the Caribbean in the wake of the warmer weather. Many species presently confined by climate to the southern U.S., such as killer bees, kudzu, hydrilla, and fire ants, will head north, too.

Of particular concern are microscopic invaders that might follow the warmth. Some would infect trees, some would hit livestock, some would damage crops. And a sobering number would afflict human beings. We could be faced with dengue fever, Chagas' disease, malaria, and other diseases we associate with more southern climes. They would be adding their weight to the microscopic exotics that already cause serious human health problems in the U.S.: bacteria, viruses, protozoa, fungi—these disease agents frequently penetrate our borders. And because we often lack immunity to non-native diseases, they are capable of inflicting more severe suffering than a native microbe to which we have already developed resistance. History shows the danger.

In 1519, Hernan Cortes and his soldiers invaded the Aztec empire, which occupied what is now central Mexico. The Spaniards employed horses, muskets, and other instruments of war that the Aztecs never had seen, but none of these new weapons came close to matching the killing power of what the Aztecs called *hueyzahuatl,* smallpox.

The Aztecs and all the other indigenous peoples of the western hemisphere constituted what epidemiologists call a naive population vis-à-vis smallpox; they were unconditioned to it and therefore extremely vulnerable. Smallpox arrived with Spanish reinforcements in 1520 and almost immediately began spreading through the Aztec population, thought to have numbered several million. Bernardino de Sahagun, a 16th-century Spanish historian, writes that the ensuing epidemic "spread over the people as great destruction. Some it quite covered on all parts—their faces, their heads, their breasts, and so on. There was a great havoc. Very many died of it. They could not walk; they only lay in their resting places and beds. They could not move; they could not stir; they could not change position, nor lie on one side; nor face down, nor on their backs. And if they stirred, much did they cry out. Great was its destruction. Covered, mantled with pustules, very many people died of them."

Ralph Bryan, medical epidemiologist for the National Center for Infectious Diseases at the Center for Disease Control and Prevention, reports that within two years smallpox killed more than half of the

Aztec population, paving the road to conquest for the Spaniards. Historian William McNeill, in his book *Plagues and Peoples*, points out that Cortes and his 600 men—advanced weaponry and native allies notwithstanding—hardly could have beaten the huge and powerful Aztec empire without the aid of smallpox. According to McNeill, those Aztecs who survived the infectious holocaust were physically, emotionally, and spiritually weakened—ripe for subjugation. He also notes the juxtaposition of Aztecs dying by the millions and Spaniards striding about in robust health. McNeill speculates that this selective slaughter convinced many Aztecs of the superiority of the Spaniard's god, undermining the Aztecs' religious convictions and making them easy targets for the Spanish friars.

Like wild fire on a windy day, smallpox burned outward from the fallen Aztec empire and from other sites of introduction, until it had swept across most of the Americas. Other non-native diseases, including measles, whooping cough, malaria, yellow fever, and diphtheria, also passed from the European explorers to the native peoples, further fueling the conflagration. With dreadful speed villages became tombs and enduring cultures fell apart. Though scholars continue to argue over the numbers, the death count most commonly has been placed in the tens of millions. One scholar estimates that 56 million native Americans died from introduced diseases during the era of European exploration—perhaps 80 to 90 percent of the total population.

Nor have we seen the final tally. European diseases have continued to afflict the few remote peoples left in the hemisphere. In 1952, measles infected 99 percent of the native people living on Ungava Bay in the far north of Quebec. Even though many received modern medical treatment, 7 percent died. In 1954, measles killed many of the rain forest dwellers deep in Xingu National Park in Brazil. During the 1990s, malaria, influenza, measles, and chicken pox have been devastating the Yanomamos, a people of the Amazon rain forest on the border between Brazil and Venezuela.

Remote indigenous peoples are not the only humans who need to worry about exotic diseases. In modern times non-native organisms have caused major epidemics among cosmopolitan populations. In 1918, the Spanish flu killed about 20 million people worldwide, including about half a million Americans. Victims often perished within three days of contracting the illness.

Three generations later, at the dawn of the twenty-first century, our rapidly growing medical wherewithal and experienced immune systems shield us better than ever. However, though we're not vulnerable to as wide a range of microbes as humans once were, we are far from safe. As Bryan said: "Immunity is not general. It's very, very specific." Meaning that we may be resistant to many more diseases than are the Yanomamos, but we're still at the mercy of unfamiliar microbes. And the likelihood that we will encounter unfamiliar diseases is increasing.

Our modern ways of living do much to promote the spread of exotic microbes. In 1992, the U.S. Institute of Medicine reported that most emerging infectious diseases stemmed from changes to our environment, such as skyrocketing populations, urbanization, and the damming of rivers. Bryan notes that such changes similarly would empower alien microbes; in fact, some of the emerging diseases that concern the Centers for Disease Control and Prevention are exotic.

Contemporary travel habits certainly contribute heavily to the dispersal of formerly local or regional diseases. The World Tourism Organization reported that in 1994, total international arrivals exceeded half a billion, enormously higher than numbers from just a few decades ago. In 1997, the University of North Carolina published a study showing one of the many ways in which human arrivals foster microbe arrivals. The researchers analyzed the waste from airplane lavatories and found that nearly half of the samples contained viruses, though supposedly the waste had been treated to kill viruses. Travel's close cousin, trade, likewise helps microbes cross natural barriers. Bryan cites the North American Free Trade Agreement and the General Agreement on Tariffs and Trade as examples of the kinds of agreements that could expand disease as they expand the global marketplace.

The speed of modern transportation matters, too, as was demonstrated as far back as the 1870s. Previously the bacillus *Yersina pestis* had been confined to the Old World, where it circulated among fleas, rats, and humans. Survivors gain immunity and the dead rapidly become uninhabitable, so this bacillus must continually find new, live hosts or it will die. In the weeks or even months needed for a sailing ship to cross the Atlantic or Pacific Oceans, *Y. pestis* would run out of susceptible victims and die out before making landfall in the New World. Then, in the 1870s, steamships changed the equation. Soon these much speedier vessels had carried live bacillus to the western hemisphere. Today *Y. pestis* has infiltrated many prairie-dog and ground-squirrel colonies in New Mexico, Colorado, California, and other western states, periodically breaking out and killing these rodents in untold numbers. Now and then it seeps into the human population. Bryan says that about 15 U.S. cases are reported each year, and despite treatment, 14 percent of these people die. As more suburbs and five-acre estates invade the territory of prairie dogs and ground squirrels, the opportunities for transmitting the bacillus rise. The disease kindled by *Yersina pestis* is what we commonly refer to as the bubonic plague.

The varied pathways by which exotic microbes can slip into the country complicate interdiction. Bryan sketched an example: "You have folks coming in from Central or South America," said Bryan, "and they bring in some steamed crabs that have not been steamed quite enough. They reheat them up here, eat them at a family gathering, and then some people come down with cholera." Cholera struck the Western Hemisphere in 1991, infecting more than one million people and killing more than 10,000. Most of the outbreak occurred in South America, but in 1992 the U.S. reported more cases of cholera than in any other year since surveillance began.

Travel and trade not only bring in diseases but the vectors for diseases, organisms that provide the link between a reservoir of infectious microbes and humans. Bryan points out that the pet trade provides some fine examples, such as the importation of rodents.

"These are wild-caught rodents from the field in the neotropics. One has to wonder whether we're importing vectors of horrible diseases." Bryan feels that current screening is inadequate. He says inspectors pretty much just eyeball rodents and the like to see if they look sick. "There's a lot of stuff coming in," said Bryan, "and there's potential for escape." Bryan also cites some unfunny monkey business. "There are some feral monkeys roaming around Florida that escaped from Tarzan movie sets back in the 1940s, so they're established. We're concerned about the potential transmission of simian herpes virus to humans, because this particular genus and species of monkey is one that can carry that virus."

One non-native vector currently of great concern to public health officials is the Asian tiger mosquito, an aggressive biter and prolific breeder. It first entered the U.S. in 1985 in Houston, apparently a stowaway in a containerized shipment of used tires from Japan. This Asian import loves used tires, breeding in the water that often pools inside them. More than two billion used tires occupy the American landscape, most near urban centers, and 250 million more are added yearly. Not surprisingly, Asian tiger mosquitoes are, as Bryan put it, "All over the bloody place." More precisely, as of 1991 they had spread to 22 states, and scientists predict they will range farther.

Those of us whose knowledge of mosquitoes is limited to scratching their bites probably think mosquitoes are pretty much alike, but they differ greatly in their abilities to transmit particular microbes. The Asian tiger mosquito ably carries LaCrosse encephalitis, dengue fever (also known as "breakbone fever"), and eastern equine encephalitis, which kills 30 percent of the people it infects. Bryan says that so far he hasn't heard of any documented case of the Asian tiger mosquito spreading these awful diseases in the U.S., but the presence of this mosquito has led to a dengue outbreak in Brazil.

How frightened should we be? Well, it's too early to panic, but concern and serious preparation seem in order, especially given the fading effectiveness of antibiotics. According to doctor and medical historian Robert Desowitz, "Unless there is a revolutionary breakthrough like that which ushered in the antibiotic age, the 21st-century doctors may be as helpless as those of the nineteenth century were in dealing with pneumonias, strep throats, diarrheas, and cuts and bruises." If these currently minor conditions could present serious problems should effective antibiotics become scarce, what of bubonic plague or dengue fever? "The birth [of a new disease] is brought about by human-made habitat changes that alter behaviors and populations of animals, pathogens, and transmitting arthropods," Desowitz observes. "If these ecological-epidemiological derangements continue unmodified, then we may be in for a bad time in the twenty-first century."

When Ralph Bryan mentioned *Outbreak*, the 1995 movie about a horrific exotic virus brought into the United States by a monkey, he surprised me by saying, "The underlying premise is not too far-fetched. If the wrong monkey got on the wrong airplane, or if the wrong person got on the wrong airplane, and that person or animal was carrying a highly lethal hemorrhagic fever virus that was trans-mitted by respiratory droplets, we'd have a disaster on our hands."

Nelson Ho took a wanted poster from the back of his truck and walked across the dirt road to a telephone pole. He firmly stapled the poster to the pole at eye level. Below the words "Wanted: Dead or Alive" were two photos showing the outlaw that Ho was tracking down, a tough hombre that goes by the name of miconia. Formally known as *Miconia calvescens*, it's an invasive tree from South America. Ho works as part of the miconia control team on the Big Island of Hawaii. I'd joined him that day as he roamed a rural subdivision called Orchid Land, about 20 miles south of Hilo. After putting up the poster, Ho returned to his truck and got several miconia leaves. They were striking: thick-veined ovals, two feet long and a foot wide, velvety dark green on top and purple on the underside. Ho stapled the leaves to the phone pole under the wanted poster: samples to show residents exactly what to watch out for.

Tahitians call miconia the "green cancer." Introduced to Tahiti in 1937 in a botanical garden, from which it escaped, miconia trees, which grow as tall as 50 feet, now dominate about two-thirds of the island. Those elephant-ear leaves create shade so dark that very few plants can grow in miconia forests. In Tahiti, the green cancer is driving 70 to 100 native plant species down the road to extinction, with ripple effects that range from erosion to disrupted invertebrate food webs. Miconia could have a similar impact in Hawaii. Bob Hobdy, the forestry manager with the Department of Land and Natural Resources, says there are a hundred-plus invasive weeds that cause serious problems in Hawaii—between 10 and 12 that are the real "bad boys"—and "then there's miconia, in a class by itself."

A botanical garden on Oahu first brought miconia to Hawaii in 1961. As early as 1971, an eminent botanist warned that miconia "is the one plant that could really destroy the native Hawaiian forest." Few listened. It wasn't until 1992, by which time miconia had spread to four islands, that authorities added it to the state noxious-weed list and nurseries reluctantly stopped selling it.

Enough residents of Hawaii finally did grasp the danger posed by miconia, however, and in the mid-1990s an anti-miconia campaign came together, involving many facets of society in the islands. Dubbed Operation Miconia, this campaign represents the largest mobilization of resources to combat an alien plant in Hawaii's history.

Nelson Ho does some of the grunt work for Operation Miconia. Putting up wanted posters is part of his job, which blends education and reconnaissance. A burly young man who wears glasses and a wisp of a black mustache, Ho has been scouting for miconia on the Big Island and warning people of its dangers since early 1997. He is a true believer who pursues his work almost too assiduously. "I found myself driving home after work," he said, "and I'd be slowing down traffic because I'd still be looking for miconia. I had to train myself not to do that." He couldn't train his unconscious mind, however. "There's a joke among our crew that you're not really part of the team until you have a dream or a nightmare about miconia. So far it's 100 percent. Everybody who's worked on it has dreamed about it." One crew member dreamed that he was cooking miconia for a meal. Ho dreamed that he was uprooting some miconia but more and more kept appearing and he couldn't keep up. Ho's nightmare is a little too close to reality on the Big Island to be entirely funny. Even as Ho surveys one area for new miconia outbreaks, known infestations in other areas continue to spread because, said Ho, "We've got more work than people to do the work."

Eradication is considered a near-certainty on Oahu and Kauai and a likelihood on Maui; that means there's a chance that miconia will be erased from all of Hawaii, if the governments of the Big Island and the state raise their level of support. If they don't, new generations of miconia will reach reproductive age, produce seeds that birds and the wind will spread, and a rare opportunity will be lost.

After Ho stapled the poster and miconia leaves to that phone pole in Orchid Land, we continued down the dirt road. Orchid Land is not your typical subdivision. Its thousands of houses are tucked away in scrubby forest on multi-acre lots, most out of sight of the road. Alien plants, including miconia, relish such disturbed, semi-wild habitat, though currently Orchid Land is not completely infested with miconia. Ho hopes to keep it that way. He drives down the road at about 20 miles an hour, scanning for that deep shade of green or that flash of purple from the underbelly of a miconia leaf. "Our prime directive is to get every adult plant before they can put out more seeds," he explains. This is critical because any tree older than four or five years will flower and produce about a million tiny seeds two or three times a year.

Ho turned into a narrow driveway and wound through the bush to a weathered two-story house. Displaying good rural manners, he remained in the car, offering the house's occupant a chance to scope out this unexpected visitor in the government truck. After half a minute a lean, hard-muscled young man clad only in cutoffs padded out to greet us. We introduced ourselves, and Ho showed the man a brochure about miconia and asked him if he'd seen any plants like that. No, he hadn't. But he'd seen the public service announcements (PSAs) about Operation Miconia on television. He was speaking in pidgin so thick I only partly understood him, but Ho didn't miss a beat. He shifted from the standard English he used with me to the pidgin he had learned growing up in Hawaii. Good communication with the locals is a key element in Operation Miconia. They need residents to spot miconia plants and report them and have set up hotlines so people can call in sightings and arrange to have the crew remove the intruders.

And people have responded. The first miconia PSA aired in 1995. Several stations ran it hundreds of times over the next two years, creating awareness among those who otherwise would never have known about the "miconia monster," as a more recent PSA labeled the invasion. Randy Bartlett, a watershed supervisor for Maui Land and Pineapple Company, told me that the first report of miconia on Kauai came from a man who saw a PSA while watching a University of Hawaii football game. Department of Agriculture officials scoured the area where the man had seen a miconia plant and found several specimens, all of which they destroyed. They've surveyed the area since and haven't found any more. "That alone was worth the PSA," said Bartlett. "It's a classic example of prevention, instead of spending millions to control it."

The benefits of educating the public about aliens goes deeper than recruiting sentinels. Wherever he goes, Ho takes his message with him. "I go to the postmaster in the little towns," he said, "and to all the little stores. I talk to the clerks, and I talk to the owners." He said that he teaches his biology students at the community college about invasives. Hawaiian school children, too, learn about the islands' invasive species and the exceptional biological resources that the aliens harm. The message is getting through. When Ho makes his rounds in places like Orchid Land, he finds that more than half of the people know about miconia. Even more important, the people are grasping the importance of the alien invasion in general. "The public is picking up on this," said Ho. "They're saying to me, 'Why don't you also go after tibouchina [a common invasive plant]?' You can take heart in that. After ten years of getting the story of Hawaii's unique biology into the public schools, getting it into the thinking of our policy makers, it's finally having an effect." Despite inadequate funding, Ho sees Operation Miconia as "a template for future efforts. It's taken a tremendous amount of blood, sweat, and tears, but I want to continue. The stakes are so high, I don't even want to contemplate losing."

Francis Oliveira and Sam Akoi are two members of the five-man crew that is clearing miconia from the mountainsides above Hana in east Maui. At the end of our drive along the Hana Highway, Bob Hobdy steered us several miles up a jolting, overgrown jeep trail to meet with the crew, which he oversees. We found Oliveira and Akoi at their truck, refilling their backpack sprayers. Both are locals, as are the other members of the crew, and they grew up playing and hunting in these mountains. That familiarity helps them cope with this hard, physically demanding job. It's difficult enough to lug heavy backpack sprayers up steep slopes and through dense undergrowth in the rain and heat, but Oliveira and Akoi say the worst part is the treacherous terrain. Hidden beneath the vegetation lies a minefield of sharp, uneven volcanic rock peppered with ankle-snapping holes. Over this landscape the crew members labor, keeping within about ten feet of each other so they won't miss a single miconia plant, even a little seedling. They pull the little ones, hack and squirt the medium ones, and girdle and spray the big ones.

So far the crew has hit about 800 acres. In the end they'll have combed through about 2,500 acres of forest, 600 of which was densely infested by miconia. This is by far the biggest infestation on Maui; the area has been overrun because it's close to the nursery that first brought miconia to Maui in the early 1970s. Hobdy figures it will take the crew about a year and a half to cover every acre. Then they have to go back over all 2,500 acres, killing resprouts and new seedlings. It's a long, tedious project, but if Oliveira, Akoi, and the rest of the crew don't wipe out this huge seed source, miconia will keep cropping up elsewhere around Maui.

The miconia project above Hana got off the ground in 1996 when some scientists and land managers went to the Maui County

Council—the island's local governing body—to tell them about the green cancer. Lloyd Loope, a researcher with the U.S. Geological Survey and a leading authority on invasive species in Hawaii, was among those scientists at the council meeting. Loope says that when the scientists and managers had finished their presentation, one of the council members looked at them and asked, "Well, why did you come here?" Smiling at the memory, Loope says that he and the other scientists and managers didn't get the councilman's drift. Puzzled, they told the councilman that they'd come to provide information about the miconia problem. According to Loope, the councilman then spelled it out, saying, "Aren't you going to ask us for money?" Now, that's something you don't hear every day, not if you're in the underfunded business of fighting invasives. Kelani English, a current council member who attended the 1996 meeting in another capacity, recalls one of the councilmen saying, "Just tell us how much you need." Laughing as he described the scene, English added, "The council had to keep hinting that they should raise the price." Clearly the council understood the importance of hitting miconia fast and hard, and they were willing to devote some of the county's scant resources to do it. In the end, the Maui Council voted unanimously to give the miconia project $104,000 for the first year. The following year the anti-invasives group requested $100,000. The council insisted that they take $132,000.

The Maui Council's eagerness to pitch in is almost unprecedented. Randy Bartlett, the Maui Pineapple watershed supervisor and an active participant in the war on miconia, feels that after a good start, Operation Miconia is starting to sputter. He applauds the Maui Council for its support of the miconia campaign, but, he said, "It's nowhere near what's needed to make it successful. And if we can't get our act together with miconia, then we may as well give up." Bartlett thinks the state needs to provide the bulk of the funding, but he characterizes their contribution so far as feeble. "Most of the [state] politicians just don't get it," said Bartlett. "We need a wave of support from the public to push the politicians." He says that many state office holders have been informed about miconia and invasive species, but they remain narrowly focused on economic issues. "They don't realize that the economy will be shot if miconia spreads," said Bartlett.

The government, at all levels, has a key role to play in opposing the alien invasion, and funding is just part of it. Individuals, neighborhoods, businesses, churches, clubs, civic groups, and nonprofit organizations can help fight the invasion in many ways. But in many cases the invasion begs for a coordinated response that these scattered entities can't provide. Miconia, for example, cannot be attacked successfully in a piecemeal fashion; it would be futile to undertake a big push to prevent miconia from establishing in Hawaii Volcanoes National Park while letting infestations prosper in adjacent areas, providing a ready seed source from which miconia could colonize the park. Government is often the best vehicle to organize the type of coordination effort needed to combat invasives.

Few voices in federal government have been calling for action, though the costly eruption of zebra mussels in the late 1980s helped the issue of exotics pick up a little steam in Washington. "Let's say it's not the sexiest subject of conversation at a cocktail party with the Grey Poupon and the white-wine set," said Conrad Burns, Republican senator from Montana. Speaking specifically of exotic range weeds, he added, "It doesn't make the headline above the fold in *The Washington Post.*" Nonetheless, concern over zebra mussels led in 1990 to the passage of the Non-indigenous Aquatic Nuisance Prevention and Control Act by Congress. This legislation raised the profile of invasive species. Federal land management agencies also helped call attention to the issue when they asked for help in their struggles with non-native species. What had long been seen as isolated local problems began to coalesce into a national issue. The alien invasion movement received a boost in 1997 when a letter signed by more than 500 scientists, agriculture officials, conservationists, and land managers urged the Administration to tackle the issue.

A Clinton Administration initiative is currently being drafted to address impacts from invasive alien species. Among other things, the initiative is expected to recommend establishment of a National Council to coordinate federal anti-invasive efforts; review existing measures and recommend new ones to be implemented by Federal agencies to reduce the risk of invasive alien species introductions and spread; prepare a comprehensive Management Plan to identify personnel and other resources needed; and encourage regional coordination including the development of new monitoring and control programs and emergency response strategies. This initiative is an important step toward the creation of a much-needed national plan of attack.

Raising the awareness level of Congress, however, will be an uphill battle. Wondering about Congress's level of interest in alien species, I spoke to an aide to George Miller, a Democratic representative from California who has long championed environmental causes. I asked her where the invasion was on Miller's agenda. "That's easy," she said, "It's not even on the radar screen. Right now we're just fighting to save the Endangered Species Act." Daniel Akaka, a Democratic senator from Hawaii, points out another problem: most members of Congress know little about the invasion. As a resident of Hawaii who recognizes the dangers all too well, Akaka has tried to educate his colleagues but with little success. "I hope[d] this bill [would] increase the awareness of the public and the government and get them to pay more attention to alien species," Akaka said speaking of legislation he introduced in 1996 that would clarify the role of APHIS.

"I think Congress is in a state of denial [about the invasion]," said Bruce Vento, a Democratic representative from Minnesota. "There are groups of people in Congress who don't want answers. They know better. They have their own sources in the form of independent, special-interest-funded groups." Vento thinks the enormous economic harm caused by alien species provides ample common ground to make the invasion a bipartisan issue, but it won't happen unless more members open their minds to new information. "It isn't all one party or the other," said Vento, "but it does seem like those in control of Congress right now aren't as interested in science as they are in political science."

"Healing comes from the Gods, and the plants are a gift," said Lyons Kapi'koho Naone III, a renowned practitioner of folk medicine. Naone has given lectures around the world, from Sweden to the South Pacific, but he is firmly rooted in his native Hawaii. He was born on Maui and he still lives on Maui. Now in his 50s, this soft-spoken man began learning about medicinal plants and the healing tradition from his grandmother when he was six years old. He has incorporated some new plants, and he dovetails his efforts with modern medicine, but the essence of his art derives from the same native plant communities that his ancestors used hundreds of years ago. But now those plant communities are being degraded by the alien invasion. "We believe in a harmony that has existed for thousands of years," said Naone. "Every plant and animal has its place, none is dominant. Humans have upset that balance."

"With so many alien species, some of them real destructive, traditional practitioners are afraid," said Naone. Tradition and the need for high-quality plants demand that Hawaiian healers gather their specimens in the wild. As of now Naone still can find what he needs, but it's getting harder. Naone also worries about the effects of alien species on native Hawaiian culture in general. "Hawaiian traditions have been gaining ground again in recent years," said Naone. "Our plants and animals are very important to those traditions. Our environment and our culture are all intertwined." Nainoa Thompson, the master navigator of the sailing canoes that carried the biting midges, shares Naone's concern. He writes, "Each time we lose another Hawaiian plant or bird or insect or forest, we lose a living part of our ancient culture. Stopping alien pests is about choosing our future and saving our past." There are plenty of practical reasons for controlling the alien invasion, but there some compelling intangible reasons as well.

Unlike Naone and Thompson, most Americans are not intimate with their native biological communities. We have become estranged from nature and have lost much of our sense of belonging to the natural world. Our sense of being part of the natural world will grow even weaker if we allow alien species to homogenize our distinctive floras and faunas. One biologist coined the term "Homogocene" to describe the new era we are entering. If this era comes to pass, there will be little left of our natural heritage except impoverished habitat and cockroaches, cheatgrass, carp, crows, and the other cosmopolitan generalists that thrive on degradation.

As a way to call attention to the threat of the Homogocene, the alien invasion could be a blessing in disguise. In order to understand the invasion, people must come to know the invasion's counterpart: our native ecosystems. And when they encounter the wonders of creation they may rediscover their ties to nature. This is happening in Hawaii, where the fight against exotic species has fostered a newfound appreciation of native species and the sublime communities in which they live.

I met Kazei Garcia in a rain forest in Wai Ka Moi Preserve on Maui. A local teenager, Kazei was one of a dozen volunteers who had sacrificed their Saturday to weed out the kahili ginger plants that are edging into this beautiful native forest. As we burrowed through the tumult of ferns and vines, yanking the occasional ginger, Kazei told me that he has been volunteering here for a couple of years; he started when his high-school ecology club signed on for one of the Saturday expeditions. Kazei worked hard. When it was time for us to move on to another spot, he lingered, ripping out just one more ginger plant, and then one more, and then one more. . . .The group leader practically had to drag him away as Kazei protested, "But there's so much!" Born on Maui, Kazei has lived there all his life. His zeal stems from his deep love for his native forests, for all his native lands. To him, that rain forest is more than just a place to hike or the watershed for his town. The rain forest is part of his home, part of who he is, part of where he belongs. That's why, he said, "I'm taking care of it."

NOTE: The asterisk (*) indicates a species
that has invaded more than one region.

ACKNOWLEDGMENTS

I welcome the opportunity to thank the people whose help made this
book possible and wish I could mention more than a handful of the
hundreds of people who shared their time and knowledge.

Mary, my wife, and Sarah, my daughter, belong at the top of the list.
They're an essential part of what gets me up and working at five in the
morning.

I'm grateful to several organizations for their support of this book,
notably the Stanley Smith Horticultural Trust, the Florida Department of
Environmental Protection, the California Exotic Pest Plant Council, the
California Native Plant Society, and Defenders of Wildlife. I'm especially
grateful to my agent, Robin Straus, to Patrice Silverstein, and to everyone
at the National Geographic Society for giving this book a good home.

Finally, I'd like to express my gratitude to a few of the scientists,
farmers, land managers, ranchers, and others who taught me about the
alien invasion. Thank you Greg Aplet, Jerry Asher, Bernd Blossey, Faith
Campbell, Liz Chornesky, Henry Coletto, Bob Ferris, Cindie Fugere,
Larry Gilbert, Doria Gordon, Janice Hill, Mark Hilliard, Alan Holt, Greg
Jubinsky, Charlene Kauhane, Lloyd Loope, Chuck Minckley, Lyons
Naone, Jenny Perry, John Randall, Craig Regelbrugge, Jake Sigg, Bill
Sladen, Jocelyn Sladen, Kassandra Stirling, Dan Thayer, Tim Tunison, Sara
Vickerman, Earl Wells, Randy Westbrooks, and Phyllis Windle.